# Energy Basics: Exploring Forms and Transformations

# ENERGY BASICS: EXPLORING FORMS AND TRANSFORMATIONS

**Prof. Dr. Mukesh Pandey**

ARCLER PRESS

www.arclerpress.com

**Energy Basics: Exploring Forms and Transformations**

*Prof. Dr. Mukesh Pandey*

**Arcler Press**
224 Shoreacres Road
Burlington, ON L7L 2H2
Canada
www.arclerpress.com
Email: orders@arclereducation.com



ISBN: 978-1-77469-938-6 (Hardcover)

# ABOUT THE AUTHOR

**Prof. Dr. Mukesh Pandey** is multi-skilled professional with around 30 years of rich exposure and versatile experience in Industries, Administration, Academics , Research domains and institution building. He received his B.E. from SATI Vidisha, M.Tech. from N.I.T. Bhopal and Ph.D. from RGPV, Bhopal. He played an instrumental role in the establishment of Rajiv Gandhi Technical University Bhopal, Madhya Pradesh since it's inception in 1999. He served as Rector(Pro Vice Chancellor), Director SoEEM & Dean of Faculty of Energy Technology. He has looked after as Coordinator for establishing IIIT, Bhopal (under PPP Mode by MHRD). He held various administrative positions as Director (R&D), Coordinator of RGPV- NBA Nodal Centre, Dy. Registrar, Member of Engineering Accreditation Evaluation Committee [EAEC] and Moderation Committee of NBA(National Board of Accreditation). Professor Pandey works in the area of Renewable Energy and Environment having focus on Solar Thermal Energy and he has developed an Energy Technology Park at RGPV, Bhopal by making use of the renewable energy resources; Solar Roof Top Plant on University Building, Energy Systems, Solar Thin Film systems, Solar wind Hybrid System, Bio-Diesel Reactor, CO2 Carbon Sequestration Unit, Dual Rotor Wind Turbine etc. Professor Pandey has an iconic track record of quality teaching, innovation and research he has been Principal Investigator of many Govt. Funded Projects by DST, MNRE, MPCST, AICTE etc including Principal Investigator of Innovative & Breakthrough Technology based 30KW CL-CSP SOLAR international R&D PROJECT (India-Japan Joint Venture Project) installed in RGPV campus. He has also facilitated as Reviewer for many national and international journals of repute and associated with many academic and research organizations at various levels. Over the past several years he has taught M.Tech. and Ph.D. Research Scholars in the areas of Energy, Environment, Direct Energy Conversion and Integrated Energy Systems and has been key-note speaker and resource person at several International and National Conferences and programmes.

# TABLE OF CONTENTS

# LIST OF FIGURES

Figure 1.10. Total US coal production, 1870–2018. (Source: By JUTIONFO - Own work, CC BY-SA 4.0, https://commons.wikimedia.org/w/index.php?curid=83572954, Creative Commons License).

Figure 1.11. Representation of U.S. oil production. (Source: By US Energy Information Administration - http://www.eia.gov/energyexplained/index.cfm?page=oil_wherehttps://www.eia.gov/energyexplained/oil-and-petroleum-products/where-our-oil-comes-from.php, Public Domain, https://commons.wikimedia.org/w/index.php?curid=27349675, Creative Commons License).

Figure 1.12. Representation of U.S. natural gas production. (Source: Manigandan P, Alam MS, Alharthi M, Khan U, Alagirisamy K, Pachiyappan D, Rehman A. Forecasting Natural Gas Production and Consumption in United States-Evidence from SARIMA and SARIMAX Models. Energies. 2021; 14(19):6021. https://doi.org/10.3390/en14196021, Creative Commons License).

Figure 2.1 Representation of the earth's radiation balance (Source: By NASA - https://web.archive.org/web/20140421050855/http://science-edu.larc.nasa.gov/energy_budget/ quoting Loeb et al., J. Clim 2009 & Trenberth et al, BAMS 2009, Public Domain, https://commons.wikimedia.org/w/index.php?curid=32285340, Creative Commons License).

Figure 2.2 Illustration of the different forms of energy (Source: BY MR SHOLTO MAUD - OWN WORK, PUBLIC DOMAIN, HTTPS://COMMONS.WIKIMEDIA.ORG/W/INDEX.PHP?CURID=83697835, Creative Commons License).

Figure 2.3. Solar energy utilization (Source: By Thompsma - Own work, CC BY-SA 3.0, https://commons.wikimedia.org/w/index.php?curid=15521895, Creative Commons License).

Figure 2.4. Illustration of the geotectonic plates (Source: Maciuk K, Apollo M, Kukulska-Kozieł A, Lewińska P. What Is the Impact of Tectonic Plate Movement on Country Size? A Long-Term Forecast. Remote Sensing. 2021; 13(23):4872. https://doi.org/10.3390/rs13234872, Creative Commons License).

Figure 2.5. Illustration of the process of photosynthesis (Source: Https://en.wikipedia.org/wiki/File:Photosynthesis_en.svg, Creative Commons License).

Figure 2.6. Representation of the types of heterotrophic nutrition (Source: Alt KW, Al-Ahmad A, Woelber JP. Nutrition and Health in Human Evolution–Past to Present. Nutrients. 2022; 14(17):3594. https://doi.org/10.3390/nu14173594, Creative Commons License).

Figure 2.7. Illustration of examples of kinetic energy (Source: Ramirez H, Le Gorrec Y. An Overview on Irreversible Port-Hamiltonian Systems. Entropy. 2022; 24(10):1478. https://doi.org/10.3390/e24101478, Creative Commons License).

Figure 2.8. Illustration of an object with mass (m) placed at height (h) from the ground (Source: Energy Education, Creative Commons License).

Figure 8.1. Role of hydrogen in global energy scenarios (Source: Dash SK, Chakraborty S, Roccotelli M, Sahu UK. Hydrogen Fuel for Future Mobility: Challenges and Future Aspects. Sustainability. 2022; 14(14):8285. https://doi.org/10.3390/su14148285, Creative Commons License).

Figure 8.2. Representation of hydrogen route versus electrical route for vehicles (Source: Ligen Y, Vrubel H, Girault HH. Mobility from Renewable Electricity: Infrastructure Comparison for Battery and Hydrogen Fuel Cell Vehicles. World Electric Vehicle Journal. 2018; 9(1):3. https://doi.org/10.3390/wevj9010003, Creative Commons License).

Figure 8.3. Illustration of the hydrogen economy (Source: Sadik-Zada ER. Political Economy of Green Hydrogen Rollout: A Global Perspective. Sustainability. 2021; 13(23):13464. https://doi.org/10.3390/su132313464, Creative Commons License).

Figure 8.4. Illustration of decarbonizing the energy system (Source: Ye Y, Dehwah AHA, Faulkner CA, Sathyanarayanan H, Lei X. A Perspective of Decarbonization Pathways in Future Buildings in the United States. Buildings. 2023; 13(4):1003. https://doi.org/10.3390/buildings13041003, Creative Commons License).

Figure 8.5. Illustration the carbon market (Source: Kim S-K, Huh J-H. Blockchain of Carbon Trading for UN Sustainable Development Goals. Sustainability. 2020; 12(10):4021. https://doi.org/10.3390/su12104021, Creative Commons License).

# LIST OF TABLES

# PREFACE

Energy is the basis of the universe, driving the dynamics of every natural phenomenon and powering the innovations that have shaped our civilization. From the gentle breeze that caresses our skin to the awe-inspiring celestial dance of stars, energy's myriad expressions surround us, yet its intricacies remain enigmatic to many. Understanding the nature of energy is crucial to comprehend the workings of our everyday lives and the grand mechanisms of the cosmos.

By 2050, the demand for energy could double or triple, as the global population rises and developing countries expand their economies. All life on earth depends on energy and the cycling of carbon. Energy is essential for economic and social development but also poses an environmental challenge.

This book comprises of chapters; the first chapter provides an introduction to energy with emphasis on the resources of energy, energy use patterns and conservation of energy. The second chapter deals with different forms of energy including kinetic energy, potential energy, thermal energy, chemical energy and nuclear energy.

The third chapter provides insights into two specific forms of energy including biological and chemical energy. This chapter discusses the energy obtained from fossil fuels and combustion. The fourth chapter discusses other two forms of energy including mechanical and electromagnetic energy and provides details of the losses and storage of these forms of energy.

The fifth chapter describes nuclear energy and discusses factors for surged interest in nuclear power along with the technology overview. The sixth chapter deals with the transformation of energy and provides insights of electric generators which convert mechanical energy into electrical energy. The seventh chapter discusses geothermal electricity which is electrical energy produced through geothermal energy sources and fuel cells which facilitates the conversion of primary fuel sources into electrical energy. The eighth chapter provides an idea about the future trends of energy.

Whether you are a student seeking a solid grasp of scientific concepts or an avid learner with a curious mind, this book provides a clear and concise exploration of energy's impact on our lives, its different forms and transformations.

***-Author***

CHAPTER

# 1 Introduction to Energy

## Contents

## INTRODUCTION

The term "energy" is a Greek compound, similar to many other abstract terms such as "hypothesis" and "sophrosyne". The term "energeia" was coined by Aristotle in his work "Metaphysics" through the combination of the Greek words "εν" (en) and "εργον" (ergon), resulting in "ενεργεια" (energeia). This concept is associated with actuality and is closely linked to "entelechia," which represents the state of complete reality. Aristotle posited that the preservation of an object's existence is contingent upon the energeia associated with its function. The term "energein" consequently acquired the connotation of movement, activity, labor, and alteration (Lewis, 2015). These definitions remained largely unchanged for almost two millennia, with no significant intellectual advancements made to refine them. It is worth noting that even prominent figures in the development of modern science held flawed understandings of energy. Over time, the term gradually acquired a level of similarity to the concepts of power and force, so distinguishing between them became challenging. David Hume expressed his dissatisfaction in 1748 in his work titled "An Enquiry Concerning Human Understanding," regarding the inherent obscurity and uncertainty associated with concepts such as power, force, energy, and necessary connection in metaphysics. Hume emphasized the constant necessity of addressing these concepts in all our intellectual investigations (Hammar et al., 2017).

During a lecture delivered at the Royal Institution in 1807, Thomas Young presented a definition of energy. It was described as the result of multiplying the mass of an object by the square of its velocity. However, it is important to note that this formulation was found to be inaccurate as it failed to account for the halving of the mass term. Furthermore, Young's definition of energy was limited solely to kinetic or mechanical energy. In 1842, the seventh edition of the Encyclopedia Britannica presented a concise and non-scientific definition of energy, characterizing it as "the power, virtue, or efficacy of a thing," albeit in a limited manner (Hong et al., 2018).

Moreover, it is employed in a figurative sense to indicate an emphasis on verbal communication. There has been a minimal alteration in the realm of public discussion since then, or more specifically, since the era of Hume, except for the increased occurrence of the term's misapplication. In the early years of the twenty-first century, the term "energy" and its related forms, such as the verb "energize" and the adjective "energetic," have become widely utilized and employed loosely to describe various lively, enthusiastic, and vigorous actions and experiences. However, there continues to be a common misconception wherein energy is often conflated with power and force. Numerous instances can be observed wherein a newly appointed influential chairman infuses renewed vitality into an established corporation; a gathering is invigorated by the commanding oratory skills of a speaker, and the realm of popular culture serves as a form of soft power for the United States (Vogelesang, 2008).

**Did You Know?**

In a single second, the Sun generates a sufficient amount of energy to fulfill the global energy demands for a period exceeding 700,000 years.

Advocates of physical fitness take an additional stance by asserting (despite the absence of logical reasoning and scientific substantiation) that they experience heightened energy levels following an especially strenuous and prolonged exercise session. The individual expresses a sense of improved well-being subsequent to the mentioned experience, which can be elucidated by a comprehensible rationale. Extended physical activity stimulates the secretion of endorphins, which are neurotransmitters that diminish the sensation of pain and elicit a state of euphoria within the brain. Consequently, this physiological response has the potential to generate a heightened sense of overall satisfaction and contentment. Engaging in a prolonged period of running can result in feelings of fatigue and, in some cases, extreme exhaustion. However, it is important to note that while such physical exertion may induce a sense of joy or even euphoria, it does not lead to an actual increase in stored energy levels compared to the pre-exercise state (Herbst et al., 2012).

## 1.1 SCIENCE OF ENERGY: ORIGINS AND ABSTRACTS

The persistent misuse of deeply rooted terminology is a prevalent phenomenon that continues to persist. However, within the realm of knowledgeable discourse, the absence of valid justifications for employing ambiguous terminology has been evident for over a century. Theoretical energy studies achieved a level of coherence and clarity, albeit not without imperfections, by the conclusion of the nineteenth century. This progress was the result of generations of cautious advancements, culminating in a significant surge of intellectual and inventive endeavors in the Western world. These endeavors established the solid groundwork for modern science and subsequently gave rise to numerous advanced concepts within the field. The foundations for these advancements were established in the seventeenth century and experienced significant progress throughout the eighteenth century. This progress was facilitated by the incorporation of Isaac Newton's comprehensive understanding of physics and the implementation of engineering experiments, notably those related to James Watt's enhancements of steam engines (Hu et al., 2020).

In the initial decades of the 19th century, Sadi Carnot, a young engineer from France, made an important advancement in the development of the contemporary comprehension of energy. Carnot formulated fundamental principles that pertain to the conversion of heat into kinetic energy, and he also established the maximum efficiency achievable by an ideal heat engine, assuming it operates reversibly (Kato et al., 2021).

Following a brief period, Justus von Liebig, a prominent figure in modern chemistry and science-based agriculture, provided a fundamentally accurate elucidation of human and animal metabolism. He attributed the production of carbon dioxide and water to food or feed oxidation (Yin et al., 2014).

The development of a fundamental law in modern physics can be traced back to a journey to Java undertaken in 1840 by Julius Robert Mayer, a young German physician who served as the ship's doctor. The blood of patients subjected to bloodletting, a medical practice that endured until the nineteenth century, exhibited a significantly higher level of brightness than those of patients residing in Germany (Granberg et al., 2021).

Mayer provided a rationale, suggesting that blood in tropical regions may exhibit lower oxidation levels than blood in temperate regions due to the reduced energy requirements for bodily metabolism in warmer climates. However, this response prompted him to inquire about another crucial query. Let's consider the phenomenon of heat loss in the tropics, specifically due to radiation. It raises the question of whether the heat lost through physical work, involving mechanical energy expenditure, also contributes to warming the surroundings. This holds true regardless of whether such work is performed in Europe or tropical Asia. Unless a mysterious origin is posited, the heat in question must also arise from the oxidation of blood (Gillman et al., 2007).

Consequently, it can be inferred that heat and work are equivalent and can be converted at a constant rate. Thus Mayer commenced the development of the principle known as the law of conservation of energy. In the year 1842, Mayer presented the initial quantitative assessment of the principle of equivalence. Subsequently, three years later, he expanded the energy conservation concept to encompass all natural occurrences, such as electricity, light, and magnetism. Mayer also provided a comprehensive account of his calculations, which were derived from an experiment involving the flow of gas between two insulated cylinders (Son et al., 2006).

James Prescott Joule, an esteemed physicist from England, successfully determined the equivalence between heat and mechanical energy. This was achieved through a series of meticulous and extensive experimental investigations. Joule employed highly sensitive thermometers to gauge the temperature of water subjected to agitation by a system of rotating vanes propelled by descending weights. This configuration facilitated a reasonably precise determination of the mechanical energy expended during the churning procedure. In 1847, Joule conducted meticulous experiments that produced a result that was found to be within a margin of less than one percent of the true value. The principle of the conservation of energy, which states that energy cannot be created or destroyed, has come to be widely recognized as the first law of thermodynamics (Li et al., 2016).

In the year 1850, Rudolf Clausius, a German theoretical physicist, authored his initial publication on the mechanical theory of heat. In this work, Clausius demonstrated that the optimal efficiency achievable by an engine employing the Carnot cycle is exclusively determined by the temperatures of the heat reservoirs, independent of the characteristics of the working substance. Furthermore, Clausius established the principle that heat cannot spontaneously flow from a colder body to a hotter body. Clausius further developed this fundamental concept and, in his 1865 publication, introduced the term "entropy" derived from the Greek word "τροπη" (meaning transformation) to quantify the level of disorder within a closed system. Clausius succinctly articulated the second law of thermodynamics, which states that the entropy of the universe tends towards a maximum value. In practical terms, it can be observed that the availability of useful energy can only diminish within a closed system, which lacks any external energy input. A lump of coal can be characterized as a high-grade, well-structured manifestation of energy possessing low entropy. When coal undergoes combustion, it generates heat, which represents a diffused, lower-grade, and disorganized manifestation of energy, exhibiting high entropy. The sequence exhibits irreversibility, as the dissipated heat (along with the emitted combustion gases) cannot be reconstructed into a solid mass of coal. Heat holds a distinctive position within the energy hierarchy, as it has the ability to convert all other forms of energy fully. However, the conversion of heat into other forms is never entirely complete, as only a fraction of the initial input is transferred to the new form (Dai et al., 2011). The second law of thermodynamics, which pertains to the universal inclination towards heat dissipation and increased disorder, has emerged as one of cosmology's most significant overarching principles. However, it is worth noting that a considerable number of individuals outside the scientific community are not well-informed about this concept. This phenomenon was notably expounded upon by C. P. Snow, a prominent English physicist, politician, and novelist, in his seminal 1959 Rede Lecture titled "The Two Cultures and the Scientific Revolution" (Stirling, 2014).

On numerous occasions, I have attended social gatherings where individuals, who are regarded as highly educated according to conventional cultural norms, have fervently expressed their disbelief regarding the lack of literacy among scientists. On a few occasions, I have experienced provocation and subsequently inquired with the company regarding the number of individuals capable of articulating the principles encompassed within the Second Law of Thermodynamics. The reaction exhibited a lack of warmth and was additionally unfavorable. I was inquiring about the scientific counterpart to the question: "Have you engaged with a literary piece authored by William Shakespeare" (Gielen et al., 2016)?

Despite its purported universality, the second law of thermodynamics seems to be consistently contravened by living organisms. These organisms exhibit a remarkable ability to conceive and develop as individuals and evolve as species and ecosystems, resulting in the emergence of highly organized and complex life forms. However, it is important to note that this matter has no inherent conflict. The application of the second law of thermodynamics is limited solely to closed systems that are in a state of thermodynamic equilibrium. The biosphere of Earth operates as an open system, continuously receiving solar energy and utilizing photosynthesis to convert it into new plant biomass. This process serves as the basis for increased order and organization, resulting in a reduction of entropy (Longfei et al., 2018).

The third law of thermodynamics, originally proposed by Walther Nernst in 1906 as the heat theorem, asserts that processes cease and entropy remains constant only when the temperature approaches absolute zero (-273 °C). In the initial decade of the twentieth century, a significant expansion of the first law of thermodynamics occurred with the groundbreaking discovery made by Albert Einstein in 1905. Einstein's conclusion posited that mass possesses an inherent connection to energy, thereby establishing a novel understanding of the physical world. As per the renowned equation E = $mc^2$, energy is determined by the multiplication of mass and the square of the speed of light. Consequently, a mere four metric tonnes of matter possess an energy content that is comparable to the global annual utilization of commercial energy. However, despite this remarkable capacity, the ability to harness the latent energy within limestone or water remains elusive (Fiamegos et al., 2020).

In commercial energy conversion, nuclear reactors are the sole means by which a modest yet significant portion of the mass is transformed into energy. Specifically, the process of fission, involving the division of nuclei, results in the liberation of energy. For instance, the fission of 1 kg of uranium 235 leads to the release of an energy quantity equivalent to 190 tonnes of crude oil. Remarkably, this energy release occurs while reducing the initial mass by a mere one gram, constituting a diminishment of one-thousandth of its original mass. On the contrary, the combustion of one kilogram of crude oil results in a minuscule reduction in the fuel's mass and the oxygen required for its combustion, amounting to a diminishment of merely one ten-billionth, rendering it impractical to measure (Xu et al., 2022).

**Did You Know?**

The earliest known use of wind energy dates back to around 5,000 BCE.

## 1.2 FUNDAMENTAL CONCEPTS

The predominant definition of energy is "the ability to perform work." However, a comprehensive understanding of this definition requires moving beyond the concept of work solely as mechanical exertion. In physics, work is the transfer of energy by applying force over a distance. In a broader sense, work can be seen as any process that induces a change, such as a change in location, speed, temperature, or composition, within a system that is affected by it. This system could be an organism, a machine, or a planet. If an individual were to remain completely still in a calm environment for ten minutes, deeply reflecting upon the statement above, no productive activity would have been undertaken, as per the conventional understanding of work, which pertains solely to exerting force in the context of a mechanical task (Kemp, 2012). However, even in physical inactivity, the human metabolism remains highly active. This is due to the utilization of energy derived from the digestion of food, which is employed in several essential processes. These processes include respiration, where oxygen is acquired, and carbon dioxide is expelled, the regulation of the body's core temperature at 37 °C, the circulation of blood, and the synthesis of various enzymes responsible for governing vital functions such as digestion and nerve signal transmission. Engaging in deep contemplation of an abstract concept does result in a slight increase in cognitive energy expenditure. However, the formation of additional neural connections within the brain during this process can be considered inconsequential. Even during periods of sleep, the brain is responsible for approximately twenty percent of the body's metabolic activity. Engaging in mentally demanding tasks only slightly increases this proportion. In the realm beyond a tranquil chamber, the diverse manifestations of energy yield a multitude of outcomes. The phenomenon of lightning, which illuminates summer skies, operates distinctively compared to a colossal harbor crane that lifts sizable steel crates from a pier and arranges them in towering stacks on a container ship. These disparities arise from a fundamental aspect of physics, namely the presence of various forms of energy and their conversions, occurring across spatial and temporal scales spanning from galactic to sub-atomic and from evolutionary to fleeting. Lightning operates within an extremely brief timeframe, resulting in the rapid illumination and heating of the atmosphere. Additionally, it facilitates the decomposition of nitrogen molecules, thereby converting the electrical energy generated from cloud-to-cloud or cloud-to-earth discharge into electromagnetic, thermal, and chemical forms of energy. On the contrary, the motors utilized in stacking cranes within container ports operate continuously, facilitating the conversion of electrical energy into mechanical energy, as well as the transformation of potential energy associated with the loaded cargo (Malen et al., 2010).

Energy is a complex and abstract concept that was introduced by physicists in the nineteenth century to encompass a wide range of natural and human-generated phenomena. It is not a singular and easily definable entity. The forms most frequently encountered include heat, motion, light, and chemical energy derived from fuels and foodstuffs. Certain conversions play a crucial role in the fundamental processes of life. For instance, photosynthesis involves the transformation of a fraction of the electromagnetic energy from light into chemical energy within bacteria and plants (Gong et al., 2012).

Additionally, the act of cooking and heating relies on the conversion of chemical energy present in biomass (such as wood, charcoal, and straw) or fossil fuels (including coals, oils, and gases) into thermal energy. The conversion of chemical energy to electrical energy in batteries facilitates the operation of numerous mobile phones, music players, and radios, thereby providing a significant level of convenience on a large scale. Certain reactions, such as gamma-neutron reactions, are considered to be relatively uncommon. These reactions involve the conversion of electromagnetic energy into nuclear energy and are primarily employed for specific scientific and industrial purposes (Sadler, 2005).

Kinetic energy is a property that is attributed to all masses in motion, regardless of their nature, such as heavy projectiles composed of depleted uranium capable of penetrating tanks or ethereal clouds rising above a tropical rainforest. The manifestations of this phenomenon are readily perceptible, and its magnitude can be easily determined through a straightforward calculation. Specifically, it is derived by taking half of the mass (m) of the object in motion and multiplying it by the square of its velocity (v), resulting in the equation $Ek = ½ mv^2$. It is important to observe that kinetic energy is directly proportional to an object's velocity square. Consequently, increasing the speed of an object by a factor of two results in a four-fold increase in energy, while tripling the speed leads to a nine-fold increase (Kannan & Vakeesan, 2016).

As a consequence, even small objects can pose significant hazards when traveling at high velocities. Tornado winds with velocities exceeding eighty meters per second, which is approximately equivalent to nearly 290 kilometers per hour, possess the capability to forcefully propel extremely lightweight fragments of straw into the solid trunks of trees. Similarly, even minuscule fragments of space debris, such as a misplaced bolt, traveling at a velocity of 8,000 meters per second, could potentially penetrate the protective suit worn by an astronaut engaged in extravehicular activity. Although the likelihood of such an event occurring has proven to be exceedingly low, it is worth noting that a space vehicle can sustain damage from a micrometeoroid traveling at an astonishing velocity of 60,000 meters per second (Han et al., 2019).

Potential energy arises due to a mass's spatial arrangement or configuration alterations. Gravitational potential energy is a pervasive phenomenon arising from alterations in an object's position within the Earth's gravitational field. This form of energy is acquired by various entities when they are elevated, including ascending water vapor, a hand raised in a gesture, a soaring bird, or a rocket ascending into the sky. The utilization of gravitational potential energy for economic benefit can be exemplified by the storage of water behind a dam, which is subsequently released to impact turbine blades and produce electricity. This widely employed method accounts for approximately twenty percent of global electricity generation. The potential energy associated with the water held in a reservoir behind a dam, or a rock in an unstable position on a slope undergoing weathering, can be determined by considering the product of the mass of the object, its average height above the ground (h), and the gravitational constant (g). This can be expressed as $Ep = mgh$. Tensioned springs, resulting from winding, serve as a prevalent illustration of the practical application of elastic potential energy. This energy is stored within the spring's deformation and subsequently harnessed as valuable work when the coil unwinds, thereby providing power to devices such as watches or walking toys (Dausch, 1992).

Biomass, encompassing living organisms such as plants, micro-organisms, animals, humans, and deceased matter found predominantly in soil, organic substances, and tree trunks, represents substantial reservoirs of chemical energy. Similarly, fossil fuels derived from the conversion of expired biomass also constitute significant repositories of chemical energy. The energy in question is stored within the atomic bonds of various biological tissues and fuels. It is subsequently released through the process of combustion, which involves rapid oxidation. This chemical reaction results in the generation of heat, thus qualifying it as an exothermic reaction. Consequently, novel chemical bonds are generated, creating carbon dioxide, commonly accompanied by the release of nitrogen and, frequently, sulfur oxides. Moreover, water is produced in the context of liquid and gaseous fuels (Granqvist et al., 2007).

Energy conversion efficiency can be defined as the proportion of desired output to the initial input. Photosynthesis can be considered an exemplification of a significantly inefficient process. Even in the case of the most productive crops, the conversion of solar radiation into new plant mass, known as phytomass, does not exceed four to five percent annually. Moreover, the global average annual rate of photosynthesis, which is often hindered by extreme cold or insufficient moisture, amounts to a modest 0.3%. When considering solely the photosynthetically active radiation, which refers to wavelengths that plant pigments and accounts can absorb for approximately forty-five percent of the incoming sunlight, the efficiency of useful energy transfer is observed to double. However, on a global scale, this efficiency remains below one percent. The phenomenon of high energy loss during a low-efficiency conversion can be understood as the limited extent to which the original energy input is successfully converted into a desired service or product. It is important to note that no energy is lost, per the first law of thermodynamics. However, as dictated by the second law of thermodynamics, a significant portion of the initial input is ultimately dissipated as unusable heat (Onile et al., 2021).

In contrast, many processes, devices, and machines exist that exhibit an efficiency exceeding ninety percent. The conversion of electricity into heat can be achieved with optimal efficiency by utilizing a baseboard resistance heater. Individuals who maintain a state of good health through the consumption of well-rounded diets can efficiently digest carbohydrates, such as sugars and starches, with a remarkable efficiency of up to 99%. Similarly, the most advanced natural gas furnaces can convert a significant portion, ranging from 95% to 97%, of the incoming fuel into heat within residential buildings. Moreover, converting electricity into the rapid rotation of large electrical motors exhibits an efficiency exceeding 95%. Conversely, the colossal turbines found in thermal power stations can convert as much as 99% of their mechanical energy into electricity as they rotate within a magnetic field (Zhu & Zhuang, 2012).

Energy phenomena exhibit a wide range of manifestations, encompassing various forms such as the intense luminosity emitted by our closest celestial body, the potential hazards associated with the release of ionizing radiation from a malfunctioning nuclear reactor, the high-temperature combustion occurring within rocket engines, and the intricately complex enzymatic reactions that take place under ordinary environmental conditions. However, measuring and quantifying all these energy phenomena is possible by employing a limited set of universally applicable units. Although traditional yardsticks continue to be used globally, contemporary scientific and engineering measurements rely on the Système International d'Unités (SI), officially established in 1960. This book will exclusively employ the appropriate International System of Units (SI). The comprehensive compilation of SI units, along with the prefixes denoting multiples and submultiples, will be provided in a subsequent section of this chapter (Haas et al., 2008).

### 1.2.1 Heat

The heat of combustion, also known as specific energy, can be defined as the disparity between the energy contained within the bonds of the first reactants and the energy contained within the bonds of the newly-formed compounds. Fuels with lower calorific values, such as wet peat and wet straw, yield less than 33% of the thermal energy generated through the combustion of gasoline or kerosene. Determining the energy content of a fuel, foodstuff, or any other combustible material can be readily achieved through the combustion of a completely dehydrated sample in a calorimeter, an apparatus designed to quantify the heat released during chemical reactions. Heat is generated through various other forms of energy conversion. Nuclear fission is a significant industrial procedure that harnesses its thermal energy for electricity generation. The heat generated from the resistance to electric current flow finds application in various domestic activities such as cooking, water boiling, and space heating (Shabgard et al., 2010).

Additionally, friction resulting from interactions within vehicle transmissions and between vehicle tires and the road generates undesirable and inevitable effects. Heat, once generated, can be transferred through three distinct mechanisms: conduction, convection, and radiation. Conduction involves the direct exchange of thermal energy through molecular contact, primarily observed in solids. Conversely, convection occurs when heat is transferred through the movement of liquids or gases. Lastly, radiation refers to the emission of electromagnetic waves by objects with a higher temperature than their surrounding environment. Most energy emitted at typical environmental temperatures originates from various sources such as the Earth's surface, plants, buildings, and human bodies. This energy is primarily in the form of invisible infrared radiation. However, objects with high temperatures exceeding 1200 °C, such as coiled tungsten wires in light bulbs, molten steel in electric arc furnaces, or distant stars, also emit radiation in the visible light spectrum (Costa et al., 1998).

The concept of latent heat refers to the quantity of energy required to induce a physical transformation without causing any alteration in temperature. For instance, the process of water into steam (known as latent heat of vaporization) at a temperature of 100 °C necessitates precisely 6.75 times more energy than the transformation of ice into water at 0 °C. The thermal energy transfer associated with water heating is responsible for the disparity observed between the gross (or higher) heating value and the net (lower) heating value of various fuels. There are two primary factors to consider in the context of energy release during fuel combustion. The first factor pertains to the overall energy output when the fuel undergoes combustion, with the condition that all the water produced is condensed into liquid form, thereby allowing for the recovery of the heat of vaporization (Agyenim et al., 2010). The second factor involves subtracting the energy necessary for the evaporation of the water generated during the combustion process. There is a disparity of approximately one percent for coke, a substance consisting predominantly of carbon, which produces solely carbon dioxide upon combustion. In the case of natural gases, the variation amounts to roughly ten percent, while pure hydrogen, which upon combustion generates solely water, exhibits a discrepancy of nearly twenty percent. The disparity in size may be more pronounced in the case of wood, although it should be noted that only a minor portion of this discrepancy can be attributed to hydrogen within the fuel. Fresh wood, characterized by its high moisture content, often exceeding seventy-five percent, poses challenges for efficient combustion. When unseasoned wood is burned, a significant portion of the thermal energy generated is utilized to evaporate the water present rather than effectively heating a room. It is important to note that if the moisture content of wood surpasses sixty-seven percent, it will not ignite (Soares et al., 2013).

## 1.3 QUANTITATIVE UNDERSTANDING: THE NECESSITY OF UNITS

The International System of Units (SI) delineates seven fundamental measures: length, mass, time, electric current, temperature, amount of substance, and luminous intensity. These units are employed directly for the measurement of the seven commonly encountered variables, as well as for the derivation of more intricate quantities. The second category encompasses a range of units that are commonly encountered in everyday scenarios, such as area, volume, density, speed, and pressure. This category also comprises more intricate concepts utilized in science and engineering, including force, energy, capacitance, and luminous flux. In order to derive the units commonly encountered in energy studies, it is sufficient to consider three fundamental variables: mass (M), length (L), and time (T). The area is denoted as $L^2$, while the volume is represented as $L^3$. The mass density is expressed as $M/L^3$, the speed as L/T, the acceleration as $L/T^2$ (which signifies the change in speed per unit of time), and the force, as per Newton's second law of motion, is denoted as $ML/T^2$ (where mass is multiplied by acceleration). Energy is consumed (work is performed) when a force is applied over a certain distance. Therefore, the dimensional energy formula can be expressed as $ML^2/T^2$ (De Ayala et al., 2021).

The scientific concept of power can be defined as measuring the energy utilization rate. Mathematically, power is expressed as the ratio of energy to time, denoted as $ML^2/T^3$. One prevalent misuse of the terminology, observed even in scholarly engineering publications, involves the conflation of power and electricity and the misrepresentation of power-generating facilities. In actuality, these facilities produce electrical energy at a fluctuating pace, contingent upon the requirements of industrial, commercial, and residential sectors for kinetic energy (generated by electric motors), thermal energy (for industrial furnaces, heat treatment, and residential heating), and electromagnetic energy (specifically its visible component, light). Moreover, it is important to note that knowing a specific machine's power rating alone does not provide any insight into its energy consumption unless the duration of its operation is known. The standard names of SI units for length, mass, and time are widely recognized by individuals. These units include the meter (m) for length, the kilogram (kg) for mass, and the second (s) for time. However, it is important to note that temperature is measured in degrees Kelvin (K) rather than Celsius (Dao et al., 2007).

Additionally, the unit of electric current is the ampere (A), the amount of substance is quantified using the mole (mol), and luminous intensity is measured in candela (cd). There exist over twenty derived units encompassing various energy-related variables, which possess distinctive names and symbols. These designations often pay tribute to prominent scientists and engineers. The Newton (N) is the unit of force, defined as $kgm/s^2$ (kilogram-meter per second squared). It represents the amount of force required to accelerate a mass of one kilogram by one meter per second. The joule (J) is defined as the product of one newton of force exerted over a distance of one meter ($kgm^2/s^2$). Power is defined as the rate at which energy is transferred or converted per unit of time, denoted as $kgm^2/s^3$. It is quantified in watts (W), where one watt equals one joule per second (J/s). Conversely, energy can be calculated by multiplying power by time. Thus, one joule is equal to one watt-second (LináWang, 2014).

**Table 1.1**. Representation of Basic SI units (Source: Vaclav Smil, Creative Commons License)

| **Basic SI Units** | | |
|---|---|---|
| **Unit of** | **Name** | **Symbol** |
| Length | Meter | m |
| Mass | Kilogram | kg |
| Time | Second | s |
| Electric current | Ampere | A |

| Temperature | Kelvin | K |
|---|---|---|
| Amount of substance | Mole | mol |
| Luminous intensity | Candela | cd |

Power density (W/m$^2$) is considered one of the most informative indicators in the field of energy research. In a more limited sense, this SI-derived unit is occasionally referred to as heat flux density or irradiance. However, it is evident that the notion of power per unit area can be applied to any energy flow, ranging from agricultural yields to the typical electricity consumption in densely populated urban regions. The measure's denominator refers to the horizontal area, which can be the Earth's surface, the footprint of a building, or any other relevant area. The power density of solar radiation that is received influences the energy flows within the biosphere (Ispas et al., 2019).

In contrast, the power density of energy consumption in households determines the rate at which fuel and electricity are utilized. Calculating the vertical power density of energy flows can provide additional insights in certain instances. This phenomenon proves to be highly advantageous in instances where intense winds, floods, and tsunamis exert substantial forces per unit of vertical area, resulting in significant devastation to vegetation and structures. Consider the catastrophic event known as the Indian Ocean tsunami that occurred on December 26, 2004 (Georgiev & Chatterjee, 2016).

One possible approach to comprehensively understanding the scale of energy and power units is by considering gravitational acceleration. At the Earth's surface, this value is approximately 9.81 m/s^2. Rounding this figure to 10 (a slight increase of less than two percent) can facilitate subsequent calculations. When a mass of one kilogram is held one meter above the ground, such as a plastic one-liter bottle of water positioned approximately at elbow height, it experiences a downward force of ten newtons due to gravitational pull. When an object with one-tenth of the bottle's mass (e.g., a small mandarin orange weighing approximately 0.1 kg) is held, it experiences a force equivalent to one newton. Lifting an orange from the kitchen floor and placing it on a counter, which is approximately one meter above the floor, necessitates the utilization of one joule of energy. If this action is completed within approximately one second, the energy expenditure will equate to a rate of one watt (Liu et al., 2021).

It is pertinent to emphasize that the power measurement does not provide any information regarding the overall quantity of energy consumed or released during the rated process. A colossal discharge of electricity, commonly known as a giant lightning bolt, possesses a power level comparable to that of the Earth's geothermal flow, estimated to be $10^{13}$ watts. However, it is important to note that the lightning bolt's existence is transient, lasting only briefly. In contrast, the geothermal flow has been persistently active since the formation of our planet approximately 4.5 billion years ago. Comparably, for individuals of smaller stature, such as a female weighing 50 kg, their basal metabolism, which pertains to converting food into usable energy for growth and physical activity, can be likened to a modest rate of approximately 60 W. This rate is akin to that of a lamp that is intermittently switched on for a limited duration of a few hours. Solar radiation is the primary and uninterrupted energy flow that influences various natural processes on Earth. This energy source, along with geothermal energy and gravitational attraction, determines the fundamental characteristics of the planet's habitable environment. Specifically, solar radiation contributes to the energy balance at a rate of $1.7 \times 10^{17}$ W, equivalent to 170 petawatts (PW). By contrast, it is noteworthy that in the year 2005, the collective utilization of all fossil fuels amounted to a worldwide rate of less than 12 terawatts (TW), which is equivalent to a mere 0.007% of the solar radiation received by the Earth (Purcell & Smith, 1976).

In the context of electricity, the measurement of current, denoted as I in equations, refers to the flow of electrons through a conductor and is quantified in units of amperes (A). André-Marie Ampère, a prominent French mathematician, played a pivotal role in the establishment of contemporary studies in the field of electrodynamics. The volt (V) is a derived unit of electrical potential, denoted as V = W/A. It is named after Alessandro Volta, a notable figure in the field of electricity who is credited with inventing the first battery. The volt serves as a quantitative measure of the electric potential disparity between a battery's positive and negative terminals. The measurement of resistance (R) experienced by an electric current is denoted in ohms and is contingent upon the conducting material as well as its physical dimensions (Sprehn et al., 2015).

Copper exhibits a significantly higher conductivity of approximately seventy percent compared to pure aluminum. Furthermore, pure aluminum demonstrates slightly over three times greater conductivity than pure iron. Additionally, it is observed that long and slender wires tend to exhibit higher resistance when compared to shorter and thicker wires. Aluminum alloys are preferred over pure copper for long-distance high-voltage lines due to their comparatively lower cost (Dunleavy et al., 2021).

In the context of electrical systems, direct current (DC) is characterized by the unidirectional flow of electrons, whereas alternating current (AC) exhibits a continuous variation in amplitude and periodically reverses its direction. Specifically, in North America, AC undergoes 120 reversals per second, corresponding to a frequency of 60 cycles per second, while in Europe, this reversal occurs 100 times per second. Ohm's law, formulated by the German mathematician and physicist Georg Simon Ohm, establishes a linear relationship between voltage, resistance, and current in direct current (DC) circuits, expressed as $V = IR$. The modification of the law governing AC circuits is necessary due to its failure to account for reactance, which represents the opposition encountered by the flow of alternating current in coils (inductive reactance) and capacitors (capacitive reactance). The modified equation, which incorporates impedance (Z), can be expressed as $I = V/Z$, where the impedance is the combined measure of reactance and resistance, denoted in ohms (Ω). However, the application of Ohm's law without adjustments is unlikely to yield significant deviations in the operation of commonplace household electrical converters, such as lights and appliances (Magnus et al., 2019).

The relationship above holds significant implications regarding electricity transmission and its safe utilization. The occurrence of electric shock and the potential for electrocution is primarily contingent upon the magnitude of the electric current that traverses an individual's body. By Ohm's law, the equation $I = V/R$ states that the current flowing through a circuit is inversely proportional to the resistance, given a constant voltage (e.g., 120 V in North America, 230 V in Europe). Consequently, an increase in resistance will result in a decrease in current. Dry skin exhibits a resistance exceeding 500 kiloohms, thereby restricting the current flow to a safe magnitude of only a few milliamperes (Bozoki & Richter, 2016).

On the other hand, it should be noted that wet skin offers a conductor with a relatively low resistance, measuring approximately 1 kΩ, thereby facilitating the flow of potentially fatal currents ranging from 100 to 300 mA. This heightened conductivity can potentially induce ventricular fibrillation, a life-threatening cardiac arrhythmia that can ultimately result in mortality. In the context of electrical safety, it is uncommon for healthy adults to experience lethality when inadvertently coming into contact with a 120 V wire using dry hands. However, the situation becomes significantly more hazardous when individuals with sweaty hands and bare feet engage with such wires on a humid summer day, even at the same voltage level of 120 V. Under equivalent circumstances, it is consistently more advantageous to make contact with an electrified conductor in Boston or Vancouver compared to London or Paris (Giorgi, 1975).

### 1.3.1 Magnitudes of Energy and Power

The fundamental units of energy and power pertain to minuscule quantities and magnitudes. One chickpea has a chemical energy content of 5,000 J, while a small vole requires 50,000 J daily for basic survival. The Honda Civic's fuel tank has a total energy capacity of approximately 1.25 billion joules (J). During operation, the fuel is consumed at a rate of eight liters per 100 kilometers, resulting in an average power output of approximately 40,000 watts (W). The winds generated during a severe thunderstorm have the capacity to release energy exceeding 100 trillion joules within a time frame of less than one hour, resulting in a power output exceeding 25 billion watts. The utilization of specific prefixes to circumvent the laborious task of writing out numerous zeros or relying on scientific notation ($10^n$) is necessary. In the realm of energy studies, in light of the diminutive nature of fundamental units, not only are the widely employed thousand-fold ($10^3$, kilo, k) and million-fold ($10^6$, mega, M) prefixes utilized, but higher multipliers such as G ($10^9$, giga), T ($10^{12}$, tera), P ($10^{15}$, peta), and E ($10^{18}$, exa) are also employed. In 1991, the International System of Units (SI) was updated to introduce additional prefixes to denote quantities of $10^{21}$ and $10^{24}$ (Zhong et al., 2023).

The most frequently employed energy multipliers are mega (MJ and GJ) and kilo (kWh), while kilo, mega, and giga (kW, MW, and GW) are commonly utilized as power multipliers. The energy content of various fuels varies, with air-dry straw containing approximately eleven MJ/kg (or GJ/t) and gasoline containing forty-four MJ/kg. The energy content of foods, which is influenced by digestibility and the portion utilized by organisms, ranges from less than one MJ/kg for leafy vegetables to nearly forty MJ/kg for pure fats. A subsequent table in this chapter provides detailed information on the average and range of energy contents for commonly used fuels and major food categories. The unit commonly employed to quantify and determine the cost of electricity consumption is the kilowatt hour (kWh), which is equivalent to one thousand watt-hours or 3.6 million watt-seconds. On average, an American household consumes approximately 1,000 kWh (1 MWh) per month. This amount can be likened to the continuous operation of fourteen 100 W lights for a duration of thirty days (Royapoor et al., 2020).

**Table 1.2.** Energy content of different fuels (Source: Vaclav Smil, Creative Commons License)

| The energy content of fuels | MJ/kg |
|---|---|
| Hydrogen | 114.0 |
| Gasoline | 44.0–45.0 |
| Crude oils | 42.0–44.0 |
| Natural gas | 33.0–37.0 |
| Anthracite | 29.0–31.0 |
| Bituminous coal | 22.0–26.0 |
| Lignite | 12.0–20.0 |
| Air-dried wood | 14.0–16.0 |
| Cereal straws | 12.0–15.0 |

Regarding power, the power ratings of small kitchen appliances, ranging from coffee grinders to coffee makers, typically fall within the range of 50-500 W. In the context of passenger cars, subcompact vehicles such as the Toyota Echo and compact vehicles like the Honda Civic exhibit power outputs of 50-95 kW, while sedans such as the Toyota Camry and Honda Accord possess power outputs ranging from 95-150 kW. Steam- and water-driven turbo-generators of considerable size possess capacities ranging from 500 to 800 MW. These generators are commonly installed in the largest fossil-fueled power plants worldwide, enabling electricity generation at rates exceeding 2 GW. The Sanxia project in China, the largest global project, is set to encompass twenty-six turbines with a combined capacity of 18.2 GW (Talkington et al., 2021).

Common metrics for power density often encompass the aggregate solar radiation received at ground level, which typically averages around 170 W/$m^2$. Additionally, the thermal energy emitted by urban centers, known as the urban heat island effect, often exceeds 50 W/$m^2$ (Adaryani et al., 2021).

On the other side of the power and energy spectrum lie the quantities that necessitate the utilization of frequently employed submultiples, namely milli (m, $10^{-3}$), micro (μ, $10^{-6}$), and nano (n, $10^{-9}$). Each keystroke made while typing this book consumes approximately 2 millijoules (mJ) of kinetic energy. Similarly, a dewdrop measuring 2 millimeters (mm) in size possesses a potential energy of 4 microjoules (μJ). Furthermore, the mass energy of a proton is estimated to be 0.15 nanojoules (nJ). The laser utilized in a CD-

ROM drive is typically specified as 5 milliwatts (mW), while a quartz watch requires approximately 1 microwatt (μW) of power. Comparatively, the power exerted by a flea during a hop is estimated to be on the order of 100 nanowatts (nW) (Pech et al., 2010).

### 1.3.2 Non-SI Units

Many artisans and engineers in English-speaking countries continue to use traditional (imperial) counterparts for all standard SI units. The energy content of fuels is frequently denoted in British thermal units (Btu), where one Btu is equivalent to 1055 joules (J). Work accomplished is often measured in foot-pounds-force (ft-lbf), with one ft-lbf being equal to 1.36 J. Power, specifically in the context of cars or motors, is commonly expressed in horsepower (hp), where one hp is equivalent to 745 watts (W). Force is typically quantified in pounds (lb), with one pound-force being equivalent to 4.45 newtons (N). Additionally, a metric unit is not derived from the seven fundamental measures of the International System of Units (SI). The calorie is defined as the quantity of thermal energy required to elevate the temperature of 1 gram of water by 1 degree Celsius, specifically from 14.5 to 15.5 °C. The caloric unit, which is slightly greater than four joules (1 cal = 4.18 J), is commonly employed in its kilocaloric form (kcal), which is a thousand times larger. An adult male who is in good health and engages in regular physical activity, with a body mass index (BMI) falling within the recommended range of 19-25, typically requires approximately 2,500 kilocalories (2.5 megacalories or 10.5 megajoules) of daily caloric intake (Kokaze et al., 2020).

However, rather than employing the appropriate scientific prefix, nutritionists started utilizing the term "Cal" (representing a large calorie) to denote a kilocalorie. This substitution was made due to the frequent misuse of the lowercase "c" instead of the uppercase letter, leading to a lack of awareness regarding the distinction between the two. It is possible that acquaintances may engage in debates asserting that a daily caloric intake of 2,500 is sufficient. Let me clarify: the specified quantity would be insufficient to sustain a twenty-gram mouse. The daily basal metabolism of a mouse, even when assuming it remains motionless for a full twenty-four hours (which is challenging for a mouse), necessitates approximately 3,800 calories (equivalent to nearly 16 kilojoules) per day. On the contrary, the daily basal metabolic rate of a healthy adult male weighing 70 kg is approximately 7.1 MJ. Engaging in activities can augment this overall energy expenditure by varying degrees, ranging from a 20% increase for individuals leading a sedentary lifestyle to a 100% increase for those involved in prolonged intense physical labor or endurance sports (Fuentes-Arderiu, 2006).

### 1.3.3 Direct and Alternating Current

The relationship between power, current, and voltage can be expressed through the equation $P = IV$, where P represents power. Additionally, voltage can be calculated as the product of current and resistance, denoted as $V = IR$. Consequently, power can also be determined by the equation $P = I^2R$. It is important to note that electricity utilization is influenced by both current and resistance. As an illustration, a white incandescent light requires a high resistance of approximately 140 Ω, while a bread toaster necessitates a relatively low resistance of about 15 Ω. The high resistance of the light would lead to the bread's combustion, whereas the toaster's low resistance only generates a reddish glow. However, a light bulb typically operates at a power consumption of 100 watts, resulting in an electrical current draw of approximately 0.8 amperes. In contrast, a toaster has a power rating of 800 watts, requiring a current draw of over 7 amperes. The equation $I^2R$ can be interpreted as follows: when transmitting an equivalent amount of power, increasing the voltage by a factor of one hundred will reduce the current by ninety-nine percent, consequently leading to a proportional decrease in resistance losses (Zaghi et al., 2010).

This elucidates the rationale behind utilizing alternating current (AC) in contemporary networks for long-distance transmission and distribution to residential areas. During the early 1880s, Thomas Edison developed the initial electric networks, which operated on direct current (DC). In this system, the voltage of the DC had to either align with the voltage required by the load, such as a light or a motor or be decreased to the appropriate level using a converter positioned in series or a resistor that dissipated the excess voltage. In order to mitigate losses in DC transmission, it would have been necessary to increase voltage and decrease current. However, this approach would have posed a significant risk of generating excessively high load voltages within residential and industrial establishments. In contrast, the transmission of high voltage alternating current (AC) over long distances with minimal losses can be effectively mitigated by utilizing transformers, which facilitate voltage reduction to acceptable levels (Cerqueira et al., 2014).

Edison exhibited resistance towards the adoption of alternating current (AC) until the year 1890, and he actively engaged in campaigning against its implementation. The technological advancements of the late 1880s, including the development of dependable transformers, AC motors, meters, and rotary DC-to-AC converters, played a crucial role in facilitating the integration of existing DC stations and networks with high-voltage AC lines. Consequently, the competition between AC and DC systems reached a definitive turning point by 1890. Although a few DC networks persisted until after World War I, it became evident that the future predominantly favored AC systems. However, a significant amount of direct current (DC) is present in electrical devices in our vicinity, either obtained through the conversion of alternating current (AC) or provided by batteries.

The superior suitability of DC motors for electric trains can be attributed to their high starting torque, which refers to the force responsible for initiating rotation (Bond, 1970). The power supply for these devices is sourced from overhead alternating current (AC) lines, which are converted to direct current (DC) onboard. Personal computers employ converters to provide a 5 V DC power output to the digital circuits. At the same time, the motors for disk drives receive a power supply of over 10 V. Various types of batteries are utilized to meet the requirements of portable DC power. These batteries are compact devices that directly convert chemical energy into electrical energy. The most prevalent type of battery found in a vast number of automobiles is the large, rechargeable lead acid battery. This battery configuration consists of six cells, collectively providing an output of 12 volts. The cathodes, which serve as the positive electrodes, are composed of lead, while the anodes, functioning as the negative electrodes, consist of lead oxide-coated lead. The space between these electrodes is filled with sulfuric acid. Car batteries provide electrical energy to starter motors and numerous small direct current (DC) motors that perform tasks previously executed manually, such as operating windows, adjusting side-view mirrors, and locking doors. The compact cylindrical batteries utilized to supply power to a growing assortment of toys, flashlights, radios, televisions, and music players are essentially enhanced iterations of the carbon-zinc cell originally devised by Georges Leclanché during the 1860s. The initial battery utilized a wet electrolyte, whereas the subsequent dry cell, which employs a slightly acidic pasty electrolyte, became more prevalent (Mansoorian et al., 2014).

Energizer's introduction of an alkaline battery in 1959 marked a significant development in battery technology. This innovation involved the replacement of the conventional carbon cathode, which utilized graphite, with manganese dioxide. Additionally, the alkaline battery incorporated a potassium hydroxide electrolyte. Cylindrical batteries universally possess a negative terminal in the form of a flat metal base, while their positive terminal is characterized by a raised metal cap. The prevalent variations include the chunky D battery, typically used for flashlights, and the finger-slim AA battery, commonly employed for electronic gadgets. These diminutive cylindrical batteries provide a voltage of 1.5 V at a capacity of 2600 Ma/h, and exhibit a relatively extended period of storage viability. However, their voltage gradually diminishes over time as they are utilized. Various types of batteries exist, such as slim lithium-ion rectangular prisms designed for laptop computers with voltage ranging from 6 to 16 V. Additionally, there are miniature silver oxide button cells with a voltage of 1.5 V, commonly utilized to power hearing aids and watches (Jiang et al., 2004).

## 1.4 ENERGY USE AND THE ENVIRONMENT

We currently reside in an era characterized by heightened consciousness regarding the environment. Politicians would encounter significant challenges in securing electoral success if they failed to articulate their commitment to environmental issues. Numerous environmental transformations have transpired over the course of four decades since the inaugural Earth Day on April 22, 1970. A selection of these alterations is delineated in the section titled "Our Earth Then and Now" (Wang et al., 2007).

The commemoration of the twenty-fifth anniversary of Earth Day in 1995 centered on advancements in air and water quality. Since 1970, there has been a significant reduction of approximately one-third in national levels of smog within the context of air pollution. In the year 1999, the city of Los Angeles did not document any ozone readings that reached the threshold required to initiate a smog alert. In contrast, two decades prior, the city experienced 120 smog alerts within a year. The level of pollution emitted per mile by new cars in 1995 was approximately 1% of that emitted by 1970 model cars. Implementing new emission standards and adopting more stringent fuel efficiency regulations resulted in a notable improvement in emission reduction during the year 2010. Since 1970, there has been a significant reduction of approximately 33% in sulfur dioxide emissions, which are widely recognized as the primary contributor to acid rain. By 1970, approximately 25% of the rivers in the United States were found to comply with the federal fishing and swimming regulations. However, this percentage increased to approximately 60% by the year 1995. These achievements were not attained without significant exertion. Expenditures allocated by federal and state entities for pollution abatement and control have experienced a significant increase since 1970. Nevertheless, apprehensions regarding federal expenditure, the national debt, and the federal government's jurisdiction persist, leading to legislative efforts aimed at substantial reforms and adjustments to environmental laws and regulations about clean air and water, hazardous waste, pesticides, endangered species, and similar matters (Dincer, 1999).

The utilization of energy resources constitutes a significant determinant impacting the natural environment. The utilization of fossil fuels has experienced a notable surge since the advent of the industrial era, resulting in a 35% elevation in the concentration of carbon dioxide within the atmosphere. This surge in fossil fuel usage has concurrently contributed to a rise in the Earth's temperature. The rising global temperature phenomenon has the potential to induce the melting of the polar ice caps, resulting in elevated ocean levels. Consequently, this could necessitate the relocation of populations residing in low-lying coastal areas. Additionally, it can signify a relocation of agricultural regions due to the northward movement of precipitation patterns, accompanied by an escalation in the prevalence of diseases (Nakata, 2004).

In addition to the utilization of energy resources, the extraction and acquisition of said resources can exert significant environmental ramifications. An illustrative example is the occurrence of an oil well blowout in the Gulf of Mexico during the month of April in the year 2010. The occurrence of a malfunction at a depth of 1 mile beneath the surface of the Gulf resulted in the uncontrolled release of crude oil into the surrounding environment for a duration of 3 months. This incident led to the discharge of approximately 200 million gallons of oil, making it the record's most significant oil spill disaster (Omer, 2008).

**Remember**

Photosynthesis, an essential biological process, facilitates the conversion of solar energy into chemical energy within plants, thereby contributing to the generation of approximately 70% of the Earth's oxygen.

The proper disposal of waste has emerged as a progressively significant environmental concern. The average daily waste production per capita in the United States amounts to approximately 4.5 pounds, resulting in an annual accumulation of over 3 tons per family. This rate of waste generation exceeds that of Europeans by a factor of two. There is a growing scarcity of suitable locations for waste materials disposal. The number of landfills has significantly decreased from 14,000 in 1970 to less than 1600 at present, despite an increase in population. The management of solid waste involves the practice of incineration, wherein the waste is subjected to combustion processes, with the resulting heat being utilized for industrial applications, electricity generation, or both. Subsequently, only the residual ash is deposited into designated landfill sites (Panwar et al., 2011).

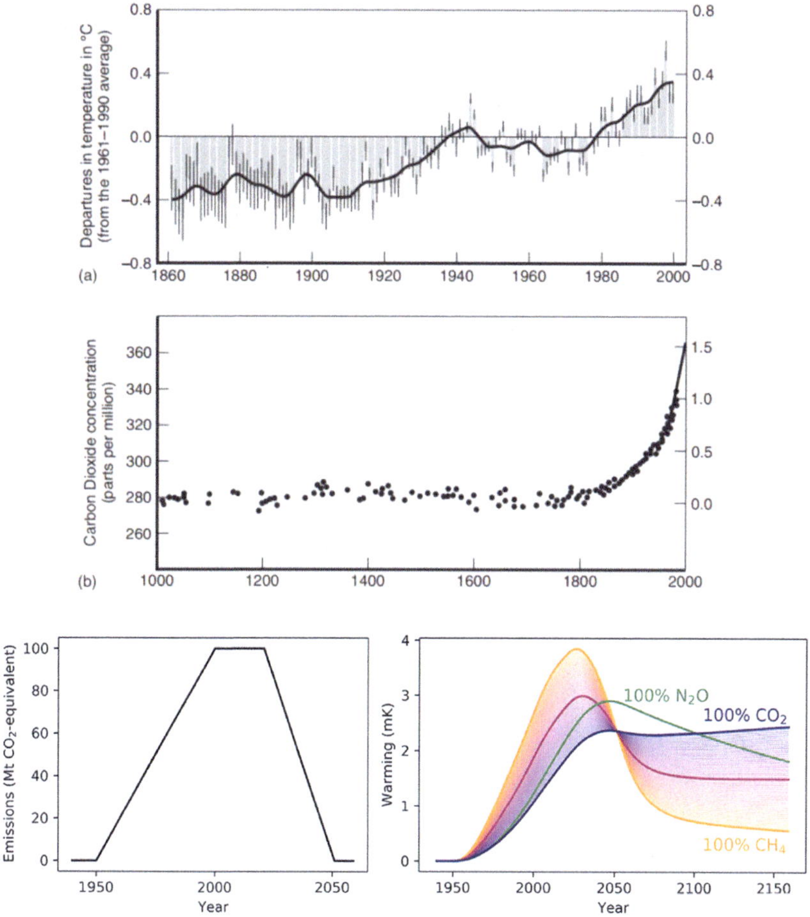


**Figure 1.1.** Correlation among global temperature change, atmospheric carbon dioxide, and methane concentrations displayed in order (Source: Lynch J, Cain M, Frame D and Pierrehumbert R (2021) Agriculture's Contribution to Climate Change and Role in Mitigation Is Distinct From Predominantly Fossil CO2-Emitting Sectors. Front. Sustain. Food Syst. 4:518039. doi: 10.3389/fsufs.2020.518039, Creative Commons License).

There is significant opposition to this particular approach due to concerns regarding potential air, water, and thermal contamination. To what extent can this problem be mitigated through recycling, the implementation of reduced packaging strategies, or other alternative approaches (Poortinga et al., 2004)?

In each of the instances above, the need to make difficult decisions arises. In order to mitigate the impact of global warming by minimizing the consumption of fossil fuels, it is imperative to explore alternative energy sources. Specifically, the promotion of solar and nuclear energy should be prioritized as potential substitutes. The present inquiry pertains to the enduring consequences arising from releasing radioactive emissions from Japan's nuclear power facilities after the catastrophic events of the March 2011 earthquake and tsunami. At what juncture do we ascertain a satisfactory level of confidence to endorse a technique for disposing radioactive waste produced at nuclear power facilities, and what alternatives can be considered substitutes for gasoline in automotive applications? Is ethanol derived from grain a viable alternative with high energy efficiency? Currently, the gasoline retailed within the United States is typically composed of 10% ethanol, predominantly derived from corn. The utilization of a 15% ethanol blend in vehicles manufactured after 2007 has been sanctioned by the Environmental Protection Agency (EPA). The question of whether food should serve as a source of energy and whether solar energy subsidies should be implemented to compete with economically cheaper fossil fuels in cases of malnourishment arises due to the finite nature of fossil fuel supplies and their detrimental environmental impact. In order to comprehensively consider the economic implications of environmental and health damage over an extended period, it is imperative to adopt a holistic approach. This entails evaluating the long-term consequences of such damage on economic systems and public health (Anker-Nilssen, 2003).

The connection between seemingly minor ethical decisions and their potentially significant consequences is frequently misunderstood. It is important to recognize that consciously making just choices, regardless of their scale, contributes to the promotion of justice on a global scale. In both personal and political contexts, individuals are ethically obligated to exercise attentiveness, refrain from succumbing to distractions, engage in sincere communication, and assume accountability for their actions, whether undertaken individually or collectively. One can choose to have faith in the prospects of the future and actively strive towards its realization and conservation, or alternatively, one can continue to navigate aimlessly, disregarding the notion that future generations will eventually inherit the consequences of our actions. The decision lies within our hands; the fate of the planet hangs in the equilibrium (Holdren et al., 2000).

## 1.5 ENERGY USE PATTERNS

From the period leading up to the 1980s, there was a consistent and substantial annual growth in global energy consumption, with particular emphasis on the United States. The graphical representation in Figure 1.5 depicts the historical energy consumption patterns in the United States spanning a period of two centuries, categorized according to the respective fuel sources utilized. From 1850 to 2000, there was a significant hundredfold increase in the utilization of commercial fuels. During the late 1940s and 1950s, there was an annual increase of 2.9% in energy consumption in the United States compared to the preceding year. During the period spanning the 1960s and early 1970s, there was a notable increase in the rate of growth, reaching a level of 4.5% annually. This growth rate would lead to a twofold increase in energy consumption within a span of 15 years. During the latter part of the 1970s, the expansion rate in energy consumption within the United States experienced a deceleration, reaching a growth rate of 3%. Subsequently, in the early 1980s, energy consumption was a notable contraction. In the year 1983, the United States exhibited an energy consumption that was 11% lower than its consumption in 1979, despite experiencing population growth. In the latter half of the 1980s, there was a modest increase in energy consumption in the United States. However, this increase was smaller in comparison to the growth rate of the Gross Domestic Product (GDP), suggesting a persistent inclination toward improved energy efficiency. During the period spanning the 1990s and early 2000s, there was a notable escalation in energy consumption, surpassing the growth rate observed in the preceding decade of the 1980s. A decline in overall energy consumption accompanied the worldwide economic recession in 2008 (Jakob et al., 2012).

In the last five decades, there has been a threefold increase in worldwide energy consumption, with projections indicating a twofold increase within the forthcoming three decades. The prevailing portion of this augmented demand in previous periods was concentrated within the industrialized nations, with fossil fuels accounting for 90% of the supply to meet this demand. Nevertheless, it is anticipated that the majority of the projected surge in energy consumption will originate from developing nations as they strive to fulfill their developmental objectives. In the present era, a staggering

1.4 billion individuals face significant challenges in obtaining or utilizing electricity, while an additional 2.7 billion individuals predominantly depend on conventional energy sources, such as biomass (e.g., wood and animal waste). These deficiencies present a significant obstacle to the advancement of social and health-related progress. According to projections, the anticipated increase in energy consumption within industrialized nations is expected to be a mere 1% annually over the course of the forthcoming decades. In contrast, developing countries are predicted to experience a growth rate of approximately 3% per year in terms of energy consumption. If these projections materialize, it is anticipated that developing nations will surpass industrialized nations in terms of energy consumption by the year 2025 (Huntington, 2010).

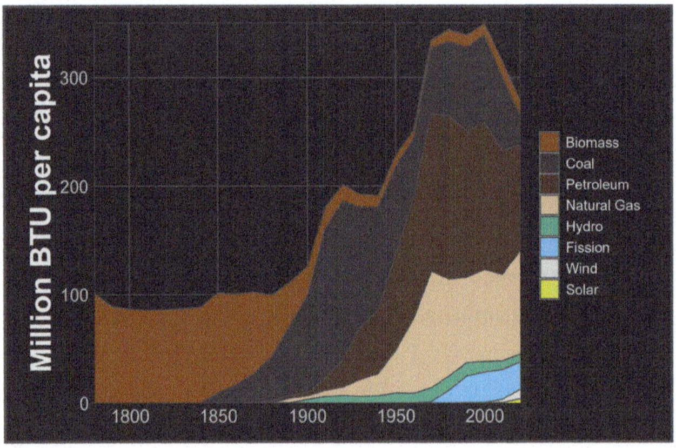


**Figure 1.2.** Energy consumption in the United States over the past 200 years, by fuel used (Source: By Wizmut - Own work, CC BY-SA 4.0, https://commons.wikimedia.org/w/index.php?curid=130481308, Creative Commons License).

Additionally, presented herein is a comprehensive elucidation of global energy consumption in 2007, categorized by geographical regions. The United States, comprising a mere 4.5% of the global population, is responsible for approximately one-fourth of worldwide energy consumption. Our nation is characterized by the rather questionable honor of possessing one of the most elevated energy consumption rates per capita among all countries. This translates to a daily energy usage equivalent to approximately 7 gallons of oil or roughly 70 pounds of coal per individual. This figure is approximately five times higher than the global average. In the event that developing nations were to raise their per capita rates to match those of developed nations, global energy consumption would experience a threefold increase. Figure

1.5 depicts the primary energy sources utilized in developed countries. The fuel mixture's composition has undergone significant alterations throughout the years. Historically, individuals augmented their own muscular strength by utilizing animals, water, and wind as mechanical energy sources for performing tasks (Heidari & Omid, 2011).

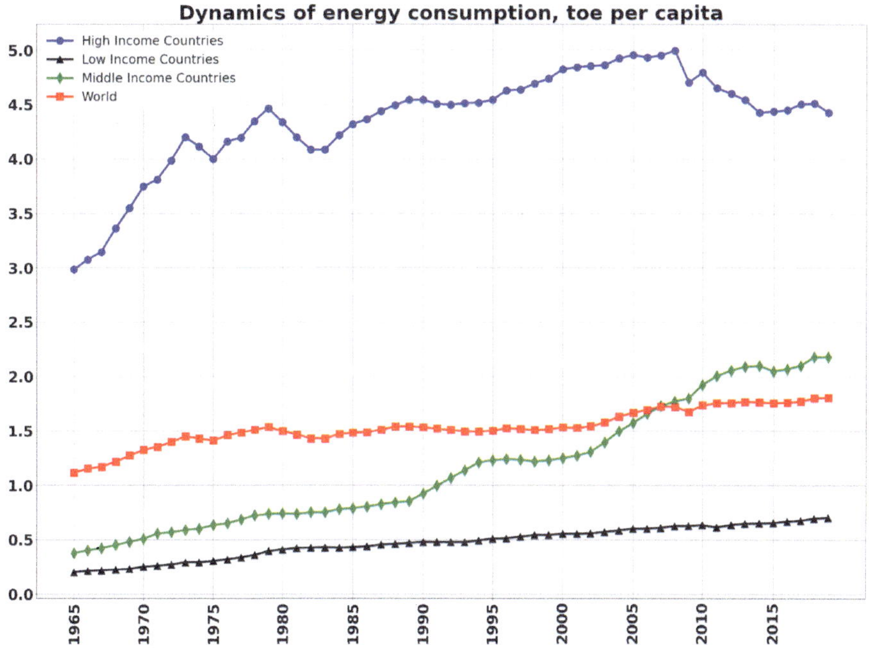


**Figure 1.3.** World energy consumption, per capita income (Source: Akaev AA, Davydova OI. A Mathematical Description of Selected Energy Transition Scenarios in the 21st Century, Intended to Realize the Main Goals of the Paris Climate Agreement. Energies. 2021; 14(9):2558. https://doi.org/10.3390/en14092558, Creative Commons License).

Pre-industrial societies exclusively relied on renewable energy sources, encompassing water, wind, solar, and biomass. These sources are characterized by their ability to be replenished indefinitely, ensuring their sustainability. The transition towards nonrenewable resources commenced during the eighteenth century, coinciding with the rise of industrialization. This period witnessed the utilization of fossil fuels for the purpose of generating steam to power steam engines, which were invented in 1763, as well as for the process of iron smelting. It is important to acknowledge that approximately 80% of our energy supply is derived from fossil fuels. Renewable energy sources, including traditional noncommercial fuels such as wood and dung, contribute approximately 20% to the global energy mix (Adegbulugbe & Akinbami, 1995).

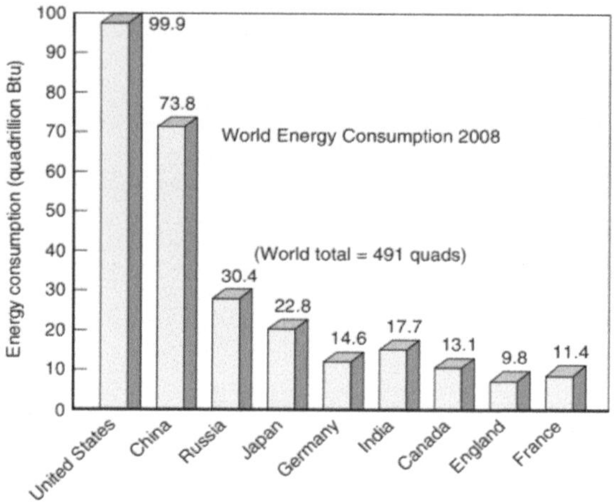


**Figure 1.4**. Illustration of world energy consumption by country (Source: John Katers, Creative Commons License).

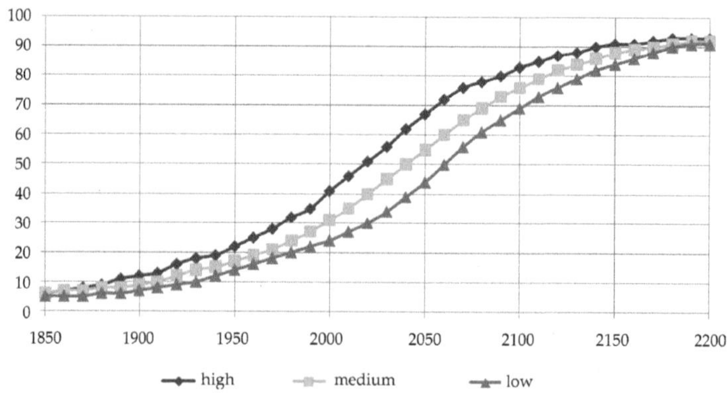


**Figure 1.5.** Representation of Energy consumption in developed countries (Source: Akaev AA, Davydova OI. A Mathematical Description of Selected Energy Transition Scenarios in the 21st Century, Intended to Realize the Main Goals of the Paris Climate Agreement. Energies. 2021; 14(9):2558. https://doi.org/10.3390/en14092558, Creative Commons License).

The inaugural instance of drilling a contemporary oil well occurred in Pennsylvania in 1859. Subsequently, the utilization of oil experienced a notable surge following the advent of the internal combustion engine in the 1870s. After 1920, there was a rapid increase in the contribution of oil, which can be attributed to the simultaneous rise in the number of engines and the availability of petroleum. The desirability of its comparatively clean-burning features was attributed to environmental considerations. Over time, the utilization of coal in industrial and power utility sectors was supplanted by the adoption of oil. Oil constitutes approximately 40% of the United States and global fuel consumption (Tso & Yau, 2003).

The utilization of natural gas within the United States remained limited and geographically concentrated until the identification of substantial reserves in Texas and Louisiana, coupled with the establishment of an extensive network of long-distance pipelines extending toward the northern regions. Currently, natural gas constitutes approximately 23% of the energy consumption in the United States, primarily for residential heating and industrial activities. The escalating identification of shale formations within the continental United States, coupled with their utilization for electricity generation, has led to a notable surge in the proportionate representation of natural gas within the overall energy sector of both the United States and the global context (Singh et al., 2002).

Within the annals of human history, the era characterized by the utilization of fossil fuels shall occupy a relatively brief temporal span. The data presented in Figure 1.6 illustrates the proportional representation of various significant energy sources in the United States throughout the course of the previous century. It is noteworthy to observe a substantial decline in the proportionate contributions of wood and coal, juxtaposed with a significant surge in oil and natural gas utilization after World War II. Before the 1940s, the United States was largely self-sufficient in oil production, meeting most of its domestic demand. Nevertheless, the United States was compelled to engage in petroleum imports starting in the latter half of the 1950s due to the simultaneous rise in energy demand and decline in domestic production (Canakci & Akinci, 2006). The pinnacle of production was attained in 1970, with a recorded output of 11 million barrels per day (MBPD). The late 1970s witnessed a notable increase in production due to the inclusion of oil from Alaska; however, the availability of this resource began to decline in 1988. Later, the aggregate production of the United States has declined to a level below 5 million barrels per day (MBPD). Figure 1.7 illustrates the production and consumption of petroleum during the latter half of the twentieth century. Following the year 1998, there was a notable increase in imports, surpassing the production level. This increase was particularly significant, as it resulted in a doubling of imports between the years 1985 and 1997. Currently, more than 50% of the oil that is consumed is imported, and it is anticipated that this proportion will continue to increase in the future. The annual expenditure on these imported goods amounts to approximately $300 billion, constituting a quarter of our trade deficit. In 2010, the United States sourced its petroleum from five primary suppliers: Canada, Mexico, Saudi Arabia, Venezuela, and Nigeria (Canakci et al., 2005).

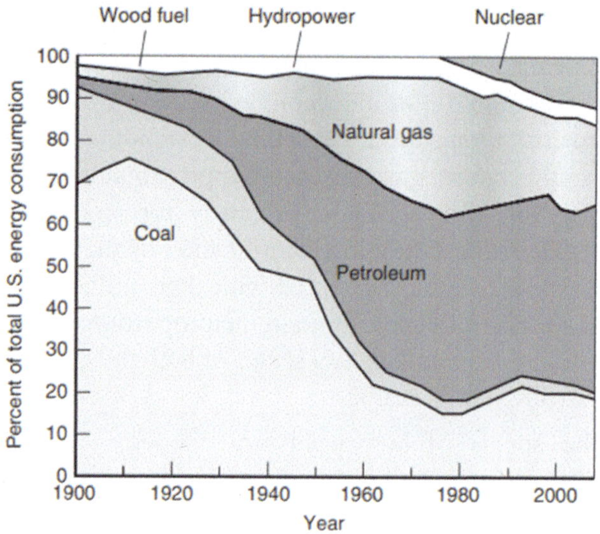


**Figure 1.6.** Illustration of U.S. energy consumption by fuel share for the past century (Source: Edward Stremer, Creative Commons License).

Renewable energy sources encompass a variety of options, such as hydroelectric power, biomass derived from wood and wood products, wind energy, photovoltaics, and illuminated solar energy, which can be utilized for heating, cooling, and electricity generation purposes. Despite accounting for less than 10% of the overall energy demand in the United States, certain technologies within this category are experiencing significant growth. Wind energy is currently experiencing rapid global growth and is considered the fastest-expanding energy source worldwide. Despite currently contributing less than 1% of the overall energy supply, the renewable energy sector is experiencing a remarkable growth rate of approximately 25% annually, with an even more impressive rate of 35% per year in Europe. Denmark's electricity generation currently relies on wind turbines for approximately 20% of its total output (Singh et al., 2003).

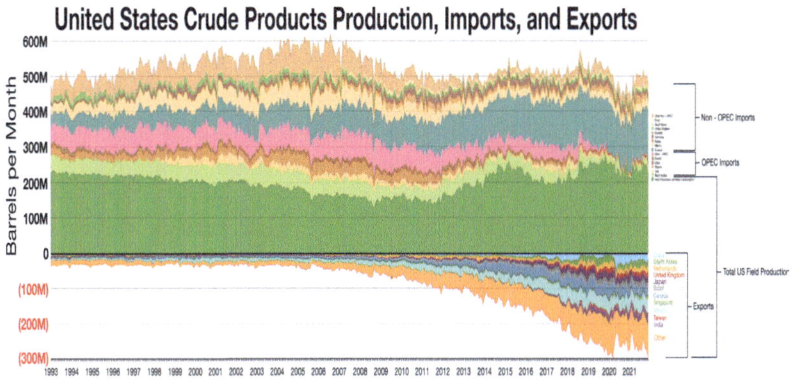


**Figure 1.7.** Illustration of U.S. petroleum production and imports: 1993–2021 (Source: By Wikideas1 - Own work https://www.eia.gov/dnav/pet/hist/LeafHandler.ashx?n=pet&s=mttfpus2&f=mhttps://www.eia.gov/dnav/pet/pet_move_expc_a_EP00_EEX_mbbl_m.htmhttps://www.eia.gov/dnav/pet/pet_move_impcus_a2_nus_ep00_im0_mbbl_m.htm, CC0, https://co, Creative Commons License).

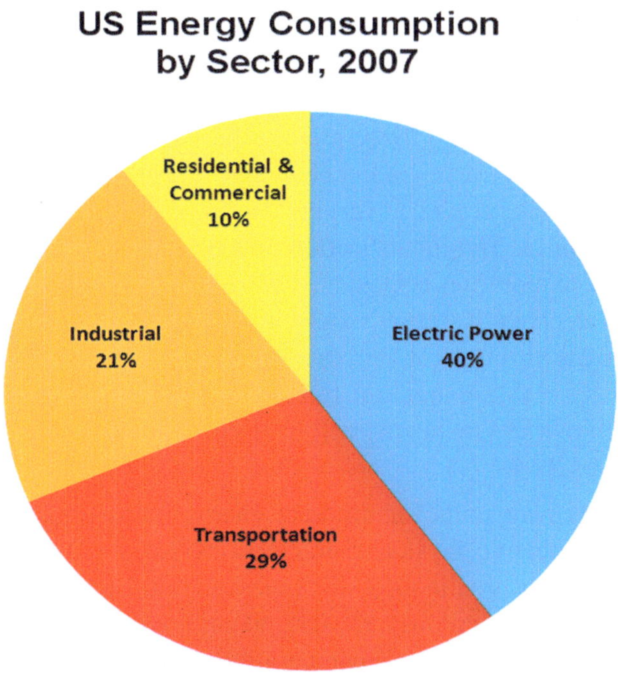


**Figure 1.8.** Representation of U.S. end uses of energy by sector (Source: Https://commons.wikimedia.org/wiki/File:US_Energy_Consumption_by_Sector_2007.PNG, Creative Commons License).

As previously mentioned in the introductory section of this chapter, it is important to note that energy is not inherently valuable, but rather its value lies in its potential for accomplishing tasks. Hence, it is imperative to scrutinize the allocation of energy. The utilization of energy is conventionally categorized into four sectors: transportation, industrial, residential (including both single and multi-family dwellings), and commercial (encompassing offices, stores, schools, etc.). The utilization of these applications in the United States in the year 2007 is depicted in Figure 1.8 (Yang et al., 2017).

## 1.6 ENERGY RESOURCES

In order to gain a comprehensive understanding of energy, it is imperative to acquire knowledge about energy resources, their inherent constraints, and their respective applications. It is imperative to possess a certain level of understanding regarding the magnitude of each energy resource and its corresponding longevity. Answering both of these inquiries presents a challenge due to the necessity of making assumptions regarding forthcoming technologies for resource extraction, future fuel pricing, and the consumption growth rate. Determining the quantities of fossil fuel resources is most straightforward for coal, as coal deposits are characterized by their widespread occurrence in extensive seams, often exposed on the Earth's surface (Chow et al., 2003). Determining the quantities of petroleum and natural gas resources poses a greater challenge due to the dispersed nature of deposits situated underground at varying depths ranging from a few hundred feet to several miles. The identification of these resources necessitates the undertaking of exploration activities. The resources above are commonly referred to as reserves. Reserves exhibit a dynamic nature as they undergo annual increments through the processes of discovery and the advancement of economically viable extraction techniques for the specific resource in question. Each of these resources will be discussed in a subsequent chapter (Ellabban et al., 2014).

The measurement units for different energy resources are determined based on their respective physical characteristics (Nordhaus et al., 1973). For instance, coal is measured in tons, oil is measured in barrels (where 1 barrel is equivalent to 42 U.S. gallons), and natural gas is measured in trillion cubic feet (tcf). In order to facilitate a comparative analysis between apples and oranges, metaphorically speaking. The unit in question is formally defined as the quantity of energy required to elevate the temperature of one pound of water by one degree Fahrenheit. One British Thermal Unit (Btu) is an approximate measure of the energy that is released through the combustion of a wooden kitchen match (Lior, 2008).

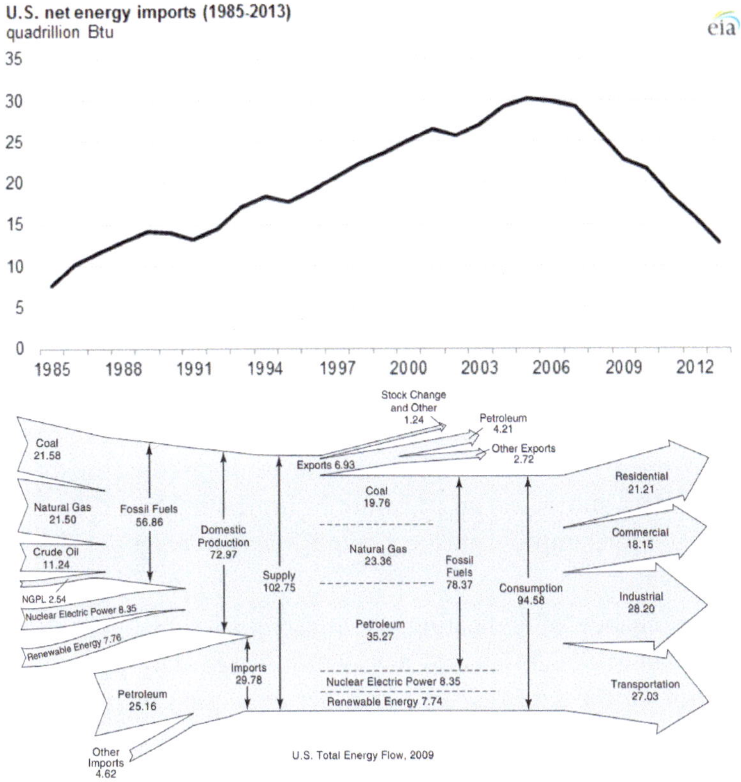


**Figure 1.9.** Illustration of U.S. total energy flow 1985-2013 (Quadrillions of Btu). (Source: By US Energy Information Administration, Today in Energy, 2 April 2014 - http://www.eia.gov/todayinenergy/detail.cfm?id=15671#, Public Domain, https://commons.wikimedia.org/w/index.php?curid=31996793, Creative Commons License).

Americans, in particular, continue to struggle with comprehending the fact that there is a scarcity of energy sources that played a crucial role in propelling the United States to its current position as a global economic powerhouse. The nation advanced by acknowledging boundless possibilities, utilizing its citizens' innovative capabilities, and the willingness to embrace risks. The economy's foundation was established with a price of $3 per barrel of oil. The current situation differs from what was previously stated. In order to sustain economic strength, individuals must recognize and acknowledge the finite nature of available resources. The lack of acknowledgment of this limited nature undeniably contributes to the historical occurrences of energy crises. The subsequent section will delve into the matter of resource depletion (Akorede et al., 2010).

**Did You Know?**

Some waste-to-energy facilities use modern incineration processes to convert trash into electricity, reducing landfill volumes and generating power.

## 1.7 EXPONENTIAL GROWTH, PEAK OIL, AND RESOURCE DEPLETION

The growth rate of consumption is a crucial determinant in estimating the lifetimes of energy resources. Previous figures presented in this chapter depicted the data above. As an illustration, from 1960 to 1970, the growth rate in energy consumption in the United States averaged 4.5% annually (Deming, 2023).

The significance of indicating the lifespan of a resource is rendered inconsequential in the absence of any mention regarding the rate at which the utilization of the said resource is either escalating or declining. There exist numerous forms of growth. However, the focus of our attention lies on exponential growth. If a quantity experiences consistent growth at a constant percentage rate over successive years, it exhibits exponential growth. Alternatively, a quantity exhibiting exponential growth consistently expands by a constant factor within a specified time interval. For instance, the duration necessary to double the quantity remains consistent irrespective of the initial amount (Brandt et al., 2013).

Let us contemplate the progression of monetary value, specifically an initial sum of $1000 deposited into a savings account that accrues a straightforward 10% annual interest rate. Table 1.3 presents the monetary balance after each year under the assumption that no funds are withdrawn. Annually, the account balance experiences a growth of 10% relative to the initial amount present at the commencement of the year. After the seventh year, the initial investment of $1000 experienced growth to $1948, representing an approximate doubling in value (Hall & Day, 2009).

**Table 1.3.** Example of 10% annual exponential growth (Source: Vaclav Smil, Creative Commons License)

| Year End | Amount ($) | Year End | **Amount ($)** |
|---|---|---|---|
| 0 | 1000 | 12 | 3138 |
| 1 | 1100 | 13 | 3453 |
| 2 | 1210 | 14 | 3798 |
| 3 | 1331 | 15 | 4178 |
| 4 | 1464 | 16 | 4596 |
| 5 | 1610 | 17 | 5056 |
| 6 | 1771 | 18 | 5562 |
| 7 | 1948 | 19 | 6118 |
| 8 | 2143 | 20 | 6730 |
| 9 | 2357 | 21 | 7403 |
| 10 | 2594 | 22 | 8143 |
| 11 | 2854 | | |

By the fourteenth year, the quantity above has experienced an approximately twofold increase, reaching approximately $3800. In the twenty-second year, $8000 was deposited in the bank, representing a twofold increase compared to the amount accessible seven years prior. The exponential growth of this quantity can be attributed to the consistent increase in the bank balance at a fixed percentage rate, resulting in a constant doubling time of approximately seven years (Pang et al., 2009).

A valuable approximate correlation exists between the doubling time, measured in years, and the percentage growth rate (Delannoy et al., 2021).

$$\text{Doubling time} \approx \frac{70 \text{ years}}{\% \text{ growth rate}}$$

Assuming an annual growth rate of 7% for electrical energy consumption, it can be projected that the quantity of electrical energy used would approximately double within a time span of 10 years. Stated differently, the required quantity of electrical power plants would exhibit a twofold increase within a decade and a fourfold increase within two decades. In order to determine the duration of a resource's lifespan, it is imperative to determine the anticipated rate of growth in its utilization concurrently (Schumack, 2006). Based on a hypothetical scenario of coal production remaining stagnant, it can be projected that the coal resources in the United States would endure for approximately five centuries. Nevertheless, assuming a 5% annual growth rate in coal production, the projected lifespan would decrease to less than seventy years (Lloyd & Subbarao, 2009).

The utilization of a specific resource will not persistently increase exponentially until its depletion, followed by an abrupt cessation. M. K. Hubbert, a researcher from the U.S. Geological Survey, has analyzed the trajectory of resource utilization, examining the patterns of growth and decline (Cavallo, 2004).

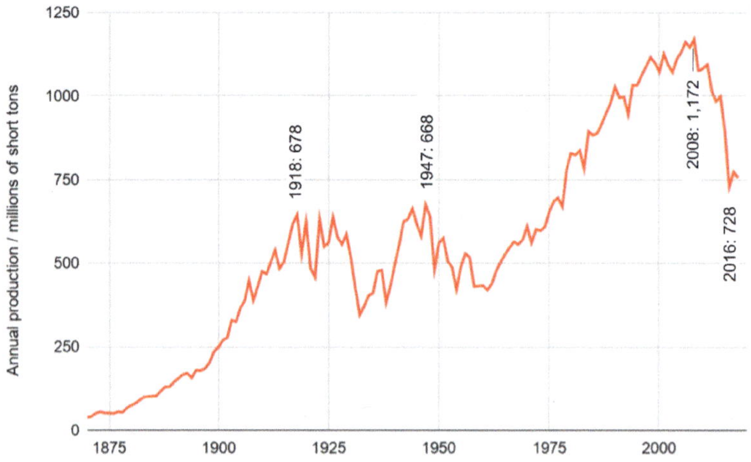


**Figure 1.10.** Total US coal production, 1870–2018. (Source: By JUTIONFO - Own work, CC BY-SA 4.0, https://commons.wikimedia.org/w/index.php?curid=83572954, Creative Commons License).

Typically, the utilization or misuse of a resource exhibits an initial phase of growth. As the depletion of high-quality deposits occurs, production attains its peak and subsequently diminishes, ultimately reaching zero upon the complete exhaustion of the resource. The production curve is expected to exhibit a bell-shaped distribution, as illustrated in Figures 1.10, 1.11, and 1.12 (Gallagher, 2011). As the depletion of a resource ensues, the processes of discovery and production become increasingly challenging, leading to a surge in prices and necessitating the substitution of alternative resources in lieu of the original fuel. When plotting the annual production as a function of time, the cumulative area beneath the curve corresponds to the aggregate quantity of the resource that is considered extractable. The quantity utilized thus far corresponds to the integral of the curve up until the current year (Hansen et al., 2017).

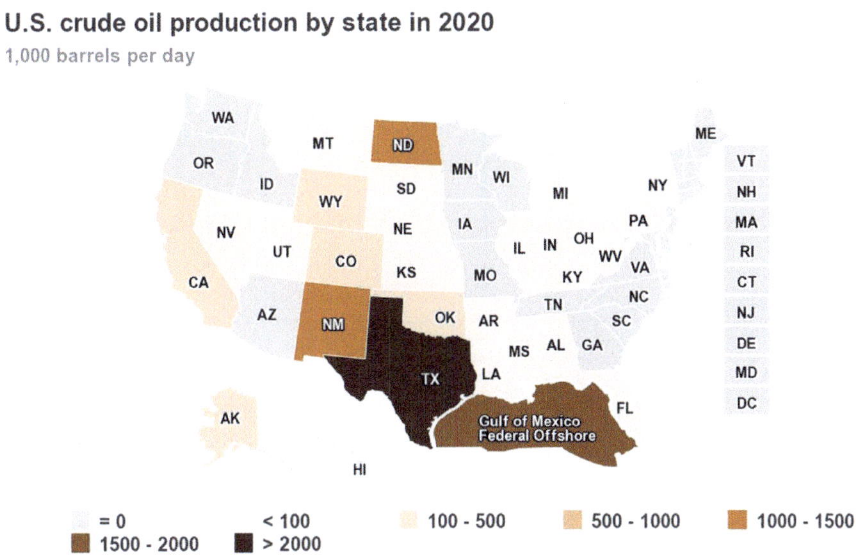


**Figure 1.11.** Representation of U.S. oil production. (Source: By US Energy Information Administration - http://www.eia.gov/energyexplained/index.cfm?page=oil_wherehttps://www.eia.gov/energyexplained/oil-and-petroleum-products/where-our-oil-comes-from.php, Public Domain, https://commons.wikimedia.org/w/index.php?curi, Creative Commons License).

Typically, the utilization or misuse of a resource exhibits an initial phase of growth. As the depletion of high-quality deposits occurs, production attains its peak and subsequently diminishes, ultimately reaching zero upon the complete exhaustion of the resource. The production curve is expected to exhibit a bell-shaped distribution, as illustrated in Figures 1.10, 1.11, and 1.12. As the depletion of a resource ensues, the processes of discovery and production become increasingly challenging, leading to a surge in prices and necessitating the substitution of alternative resources in lieu of the original fuel. When plotting the annual production as a function of time, the cumulative area beneath the curve corresponds to the aggregate quantity of the resource that is considered extractable. The quantity utilized thus far corresponds to the integral of the curve up until the current year (Patzek, 2008).

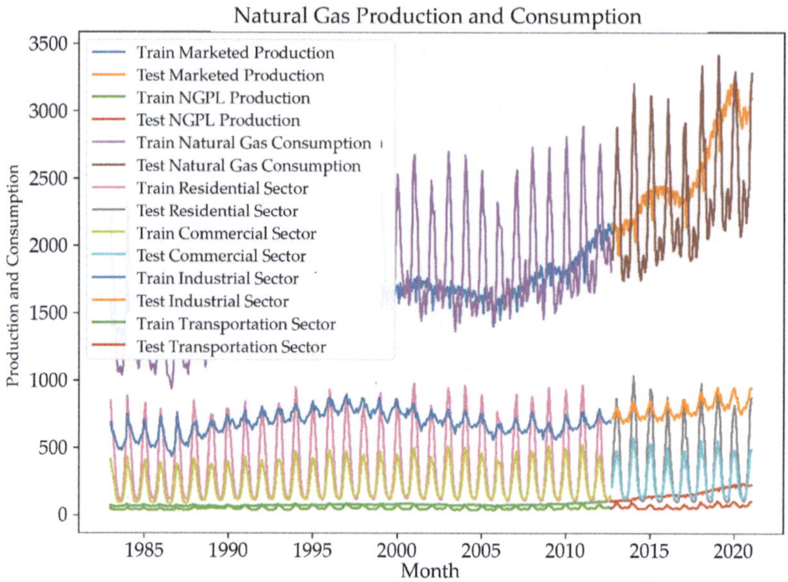


**Figure 1.12.** Representation of U.S. natural gas production. (Source: Manigandan P, Alam MS, Alharthi M, Khan U, Alagirisamy K, Pachiyappan D, Rehman A. Forecasting Natural Gas Production and Consumption in United States-Evidence from SARIMA and SARIMAX Models. Energies. 2021; 14(19):6021. https://doi.org/10.3390/en14196021, Creative Commons License).

The utilization of bell-shaped production curves enables the estimation of the timeframe until the total exhaustion of a given resource. Additionally, these curves estimate the time at which maximum production will be achieved. The graph depicted in Figure 1.10 illustrates the US coal production curve (Perifanis, 2022). Based on the graph, it can be inferred that the coal reserves exhibit a substantial magnitude, projected to endure for over five centuries at the present rate of consumption. Furthermore, the graph indicates that the pinnacle of coal production is anticipated to transpire nearly two centuries from now. The circumstances exhibit notable disparities when it comes to natural gas and petroleum. Figure 1.9 depicts the production of oil in the United States. The data indicates that the projected oil production rate in the United States will decrease by approximately one-third within a span of 20 years (Méplan & Gignoux, 1996).

Additionally, it suggests that the zenith of oil production in the United States was expected to transpire around 1970, a projection that indeed materialized. The phenomenon characterized by the highest oil production level is commonly called "peak oil" (Brecha, 2013). The current situation does not necessarily imply a depletion of oil reserves but rather signifies the culmination of the era of easily accessible and inexpensive oil. Furthermore, the paper provides a comprehensive analysis of the issue above and delves into an in-depth examination of the production curve about global oil (Celestin & Pasko, 2011). The observed pattern resembles that observed in the United States, albeit with the distinction that the projected apex is anticipated to transpire within the timeframe spanning from the current period to 2035.

There is a noticeable contrast. The input into this curve encompasses both political and geological factors. Similar observations can be derived from Figure 1.10 regarding natural gas and its estimated production curve, suggesting that U.S. gas production peaked in 1973. However, the natural gas production rate has undergone significant changes in the early years of the twenty-first century (Shah et al., 2010). This can be attributed to the discovery of substantial shale gas deposits within the continental United States, offshore deposits, and advanced drilling techniques. These developments have effectively resulted in doubling the estimated amount of recoverable gas resources in the US.

The proportion of natural gas imports concerning our overall gas consumption experienced a decline to approximately 10% in the year 2010. Could events like those affecting gas reserves significantly alter our current production trajectory for oil? According to the perspective of numerous experts, there is a consensus that the global oil supply has already reached its peak, with the depletion of economically viable oil reserves occurring rapidly (Hashimoto et al., 2020).

## 1.8 ENERGY CONSERVATION

The aggregate energy consumption associated with a given activity can be conceptualized as comprising two primary components: the efficiency of the activity, which quantifies the ratio of useful energy output to the energy input, and the duration or quantity of the activity, such as the distance traveled or the duration of time the lights were in operation (Blough & Santi, 2002). The monthly energy consumption of a vehicle is contingent upon two factors: the car's fuel efficiency, measured in miles per gallon, and the total distance traveled during that month. Efforts about energy conservation typically focus on either one or both of these factors. This document enumerates two distinct methodologies (Velloso & Jardim, 2006):

### 1.8.1 Technical Fix

This entails the utilization of fuel in a more efficient manner to accomplish a given task, such as operating a vehicle equipped with a more effective engine (thereby reducing the energy demand associated with said activity) (Cheng et al., 2021).

### 1.8.2 Lifestyle Change

This entails the deliberate utilization of a reduced amount of fuel through actions such as adjusting the thermostat to a lower setting or decreasing the number of miles driven, thereby diminishing the frequency of the respective activities (Bertoldi, 2022).

The fundamental principles of physics inherently constrain the efficacy of technical solutions for energy conservation. Nevertheless, there is ample scope for enhancing this energy conservation approach, particularly optimizing energy efficiency for specific tasks. As an illustration, a fluorescent bulb with a power rating of 20 watts produces an equivalent amount of illumination as a 75-watt fluorescent bulb while also exhibiting a tenfold longer lifespan. Although the upfront cost of a fluorescent bulb may be higher, the subsequent savings in electricity expenses throughout a year's typical usage will ultimately offset the initial investment. Replacing incandescent bulbs with fluorescent or LED alternatives would reduce demand for electric power plants. The capital required for establishing a facility dedicated to manufacturing energy-efficient light bulbs is significantly lower compared to the investment needed for constructing an electrical power plant. This particular branch of economics holds significant significance in all nations (Maio et al., 2007).

In energy conservation, the complexities extend beyond technological factors, as the energy consumption rate is also contingent upon the frequency of the respective activity. Numerous obstacles impede the implementation of the measures that will be examined in this book, including market limitations, such as the upfront expenses associated with home insulation. There is a prevailing hesitancy to transition towards what are perceived as "lifestyle changes," such as alterations in comfort regulation and preferences for specific materials (Young, 1991).

There is a prevalent viewpoint among individuals that energy prices ought to accurately reflect the anticipated expenses associated with replenishing the depleting reserves of nonrenewable energy sources, such as oil and natural gas, instead of solely reflecting the costs of their extraction. Societies are unlikely to transition towards renewable energy technologies and more efficient machinery if fossil fuels are priced significantly lower. The primary factor contributing to the decline in per capita energy consumption in the early 1980s was the significant increase in oil prices. Before that period, per capita energy consumption had a consistent upward trend. From 1900 to 1980, there was a notable increase in per capita energy consumption, which escalated to 320 million British thermal units (Btu) annually. To what extent did the quality of life experience significant improvement? Do you believe you are four times more prosperous than your great-grandparents? The per capita electricity consumption experienced a six-fold increase between 1950 and 2000 (Ding et al., 2017).

- Compelling arguments support the increasing focus on energy conservation action technologies serve as economically viable alternatives to implementing supplementary supply technologies. In most instances, the cost of preserving a unit of oil is lower than the expenditure involved in creating a fresh unit of an alternative oil source. According to the International Energy Agency's statement in 1987, allocating resources towards energy conservation yields a more favorable return than investment in energy supply.
- Conservation efforts have the potential to effectively utilize the earth's finite energy resources, benefiting not only the US as a whole but also other nations. Most less developed nations imported oil to meet at least 75% of their commercial energy requirements. The implementation of conservation practices will allow for a sufficient timeframe to explore and potentially harness renewable resources, such as solar energy, which have the potential to be indefinitely sustainable.
- The implementation of conservation practices has the potential to mitigate environmental pollution. By reducing energy consumption, the potential consequences include decreased levels of air and water pollution, reduced instances of radiation and thermal pollution, as well as diminished occurrences of global warming and acid rain (Hara et al., 2015).
- The implementation of conservation methods can be expedited faster than the pace at which we can enhance the availability of resources. The process of establishing a new coal mine typically requires a time frame ranging from 2 to 4 years. Similarly, the

construction of a gas-turbine power plant typically takes around 2 to 3 years. In contrast, the development of a coal-fired electric generating plant generally spans a period of 5 to 7 years. Lastly, the construction of a nuclear power plant is a significantly more time-consuming endeavor, typically requiring a duration of 9 to 11 years. Numerous conservation measures can be promptly initiated due to the accessibility and simplicity of existing technologies, exemplified by the implementation of enhanced building insulation. The financial requirements for implementing energy-saving measures are lower compared to capital-intensive supply technologies.

- The preservation of fossil-fuel resources holds significant importance for the future due to their greater significance as raw materials for chemical industries, such as pharmaceuticals and plastics, compared to their utilization as fuels in power generation.
- Individuals can easily implement conservation techniques, offering financial and energy-saving incentives. These practices can also contribute to personal health. For instance, bicycling offers a greater level of physical exercise than driving a car (Dillman et al., 1983).

## SUMMARY

Energy is an intrinsic characteristic of matter that possesses the capacity to perform work or induce alterations. The phenomenon is found in diverse manifestations and has the capacity to transform one manifestation into another. The joule is the standard unit of energy in the International System of Units (SI), with other frequently used units including the calorie and kilowatt-hour. Energy can be derived from diverse sources, encompassing fossil fuels, renewable sources, and nuclear reactions. Energy conservation encompasses the practice of optimizing energy utilization and minimizing superfluous energy dissipation. The factor above assumes a pivotal role in the implementation of sustainable energy practices. The generation and utilization of energy can yield substantial environmental consequences, including the emission of greenhouse gases, the contamination of air, and the depletion of natural resources. There exist multiple manifestations of energy, encompassing mechanical energy, thermal energy, electromagnetic energy, chemical energy, and nuclear energy.

# REFERENCES


1. Adaryani, M. R., Taher, S. A., & Guerrero, J. M. (2021). Improved direct model predictive control for variable magnitude variable frequency wave energy converter connected to constant power load. Journal of Energy Storage, 43, 103175.
2. Adegbulugbe, A. O., & Akinbami, J. F. K. (1995, May). Urban household energy use patterns in Nigeria. In Natural Resources Forum. Oxford, UK: Blackwell Publishing Ltd. (Vol. 19, pp. 125-132).
3. Agyenim, F., Hewitt, N., Eames, P., & Smyth, M. (2010). A review of materials, heat transfer and phase change problem formulation for latent heat thermal energy storage systems (LHTESS). Renewable and Sustainable Energy Reviews, 14(2), 615-628.
4. Akorede, M. F., Hizam, H., & Pouresmaeil, E. (2010). Distributed energy resources and benefits to the environment. Renewable and Sustainable Energy Reviews, 14(2), 724-734.
5. Anker-Nilssen, P. (2003). Household energy use and the environment—a conflicting issue. Applied Energy, 76(1-3), 189-196.
6. Bertoldi, P. (2022). Policies for energy conservation and sufficiency: Review of existing policies and recommendations for new and effective policies in OECD countries. Energy and Buildings, 264, 112075.
7. Blough, D. M., & Santi, P. (2002, September). Investigating upper bounds on network lifetime extension for cell-based energy conservation techniques in stationary ad hoc networks. In Proceedings of the 8th annual international conference on Mobile computing and networking (Vol. 3, pp. 183-192).
8. Bond, A. M. (1970). Direct current, alternating current, rapid, and inverse polarographic methods for determination of tin (IV). Analytical Chemistry, 42(11), 1165-1168.
9. Bozoki, E., & Richter, M. (2016). Entrepreneurship, Institutions and Economic Growth: A quantitative study about the moderating effects of institutional dimensions on the relationship of necessity-and opportunity motivated entrepreneurship and economic growth.51(43), 25741-25745.
10. Brandt, A. R., Millard-Ball, A., Ganser, M., & Gorelick, S. M. (2013). Peak oil demand: the role of fuel efficiency and alternative fuels in a global oil production decline. Environmental science & technology, 47(14), 8031-8041.
11. Brecha, R. J. (2013). Ten reasons to take peak oil seriously. Sustainability, 5(2), 664-694.

12. Canakci, M. U. R. A. D., & Akinci, I. (2006). Energy use pattern analyses of greenhouse vegetable production. Energy, 31(8-9), 1243-1256.
13. Canakci, M. U. R. A. D., Topakci, M. E. H. M. E. T., Akinci, I., & Ozmerzi, A. (2005). Energy use pattern of some field crops and vegetable production: Case study for Antalya Region, Turkey. Energy conversion and Management, 46(4), 655-666.
14. Cavallo, A. J. (2004). Hubbert's petroleum production model: an evaluation and implications for world oil production forecasts. Natural resources research, 13, 211-221.
15. Celestin, S., & Pasko, V. P. (2011). Energy and fluxes of thermal runaway electrons produced by exponential growth of streamers during the stepping of lightning leaders and in transient luminous events. Journal of Geophysical Research: Space Physics, 116(3), 2144-2145.
16. Cerqueira, A. A., Souza, P. S. A., & Marques, M. R. C. (2014). Effects of direct and alternating current on the treatment of oily water in an electroflocculation process. Brazilian Journal of Chemical Engineering, 31, 693-701.
17. Cheng, Z., Li, L., & Liu, J. (2021). Research on China's industrial green biased technological progress and its energy conservation and emission reduction effects. Energy Efficiency, 14(5), 42.
18. Chow, J., Kopp, R. J., & Portney, P. R. (2003). Energy resources and global development. Science, 302(5650), 1528-1531.
19. Costa, M., Buddhi, D., & Oliva, A. (1998). Numerical simulation of a latent heat thermal energy storage system with enhanced heat conduction. Energy conversion and management, 39(3-4), 319-330.
20. Dai, Y., Liu, W., Formo, E., Sun, Y., & Xia, Y. (2011). Ceramic nanofibers fabricated by electrospinning and their applications in catalysis, environmental science, and energy technology. Polymers for Advanced Technologies, 22(3), 326-338.
21. Dao, M., Lu, L., Asaro, R. J., De Hosson, J. T. M., & Ma, E. (2007). Toward a quantitative understanding of mechanical behavior of nanocrystalline metals. Acta Materialia, 55(12), 4041-4065.
22. Dausch, J. G. (1992). The problem of obesity: fundamental concepts of energy metabolism gone awry. Critical Reviews in Food Science & Nutrition, 31(4), 271-298.
23. De Ayala, A., Foudi, S., Solà, M. D. M., López-Bernabé, E., & Galarraga, I. (2021). Consumers' preferences regarding energy efficiency: a qualitative analysis based on the household and services sectors in Spain. Energy Efficiency, 14(1), 3.

24. Delannoy, L., Longaretti, P. Y., Murphy, D. J., & Prados, E. (2021). Peak oil and the low-carbon energy transition: A net-energy perspective. Applied Energy, 304, 117843
25. Deming, D. (2023). M. King Hubbert and the rise and fall of peak oil theory. AAPG Bulletin, 107(6), 851-861.
26. Dillman, D. A., Rosa, E. A., & Dillman, J. J. (1983). Lifestyle and home energy conservation in the United States: the poor accept lifestyle cutbacks while the wealthy invest in conservation. Journal of Economic Psychology, 3(3-4), 299-315.
27. Dincer, I. (1999). Environmental impacts of energy. Energy policy, 27(14), 845-854.
28. Ding, Q., Cai, W., Wang, C., & Sanwal, M. (2017). The relationships between household consumption activities and energy consumption in China—An input-output analysis from the lifestyle perspective. Applied Energy, 207, 520-532.
29. Dunleavy, L., Preston, N., Bajwah, S., Bradshaw, A., Cripps, R., Fraser, L. K., ... & Walshe, C. (2021). 'Necessity is the mother of invention': Specialist palliative care service innovation and practice change in response to COVID-19. Results from a multinational survey (CovPall). Palliative Medicine, 35(5), 814-829.
30. Ellabban, O., Abu-Rub, H., & Blaabjerg, F. (2014). Renewable energy resources: Current status, future prospects and their enabling technology. Renewable and Sustainable Energy Reviews, 39, 748-764.
31. Fiamegos, Y., Dumitrascu, C., Ghidotti, M., & De la Calle Guntinas, M. B. (2020). Use of energy-dispersive X-ray fluorescence combined with chemometric modelling to classify honey according to botanical variety and geographical origin. Analytical and bioanalytical chemistry, 412, 463-472.
32. Fuentes-Arderiu, X. (2006). Biological variation of non-SI traceable biological quantities: example of proteins. Clinical Chemistry and Laboratory Medicine (CCLM), 44(12), 1497-1497.
33. Gallagher, B. (2011). Peak oil analyzed with a logistic function and idealized Hubbert curve. Energy Policy, 39(2), 790-802.
34. Georgiev, G. Y., & Chatterjee, A. (2016). The road to a measurable quantitative understanding of self-organization and evolution. Evolution and transitions in complexity: the science of hierarchical organization in nature, 4, 223-230.
35. Gielen, D., Boshell, F., & Saygin, D. (2016). Climate and energy challenges for materials science. Nature Materials, 15(2), 117-120.

36. Gillman, M. W., Barker, D., Bier, D., Cagampang, F., Challis, J., Fall, C., ... & Thornburg, K. L. (2007). Meeting report on the 3rd International Congress on Developmental Origins of Health and Disease (DOHaD). Pediatric Research, 61(5), 625-629.
37. Giorgi, A. (1975). Convergence and divergence of qualitative and quantitative methods in psychology. Duquesne studies in phenomenological psychology, 2, 72-79.
38. Gong, J., Liang, J., & Sumathy, K. (2012). Review on dye-sensitized solar cells (DSSCs): Fundamental concepts and novel materials. Renewable and Sustainable Energy Reviews, 16(8), 5848-5860.
39. Granberg, F., Wang, X., Chen, D., Jin, K., Wang, Y., Bei, H., ... & Djurabekova, F. (2021). Origin of increased helium density inside bubbles in Ni (1− x) Fex alloys. Scripta Materialia, 191, 1-6.
40. Granqvist, C. G., Niklasson, G. A., & Azens, A. (2007). Electrochromics: fundamentals and energy-related applications of oxide-based devices. Applied Physics A, 89(1), 29-35.
41. Haas, R., Nakicenovic, N., Ajanovic, A., Faber, T., Kranzl, L., Müller, A., & Resch, G. (2008). Towards sustainability of energy systems: A primer on how to apply the concept of energy services to identify necessary trends and policies. Energy Policy, 36(11), 4012-4021.
42. Hall, C. A., & Day, J. W. (2009). Revisiting the Limits to Growth After Peak Oil: In the 1970s a rising world population and the finite resources available to support it were hot topics. Interest faded—but it's time to take another look. American Scientist, 97(3), 230-237.
43. Hammar, L., Gullström, M., Dahlgren, T. G., Asplund, M. E., Goncalves, I. B., & Molander, S. (2017). Introducing ocean energy industries to a busy marine environment. Renewable and Sustainable Energy Reviews, 74, 178-185.
44. Han, J., Vahidi, A., & Sciarretta, A. (2019). Fundamentals of energy efficient driving for combustion engine and electric vehicles: An optimal control perspective. Automatica, 103, 558-572.
45. Hansen, J. P., Narbel, P. A., & Aksnes, D. L. (2017). Limits to growth in the renewable energy sector. Renewable and Sustainable Energy Reviews, 70, 769-774.
46. Hara, K., Uwasu, M., Kishita, Y., & Takeda, H. (2015). Determinant factors of residential consumption and perception of energy conservation: Time-series analysis by large-scale questionnaire in Suita, Japan. Energy Policy, 87, 240-249.
47. Hashimoto, K., Huh, K. B., Kim, K. Y., & Watanabe, R. (2020). Exponential

growth of out-of-time-order correlator without chaos: inverted harmonic oscillator. Journal of High Energy Physics, 2020(11), 1-25.

48. Heidari, M. D., & Omid, M. (2011). Energy use patterns and econometric models of major greenhouse vegetable productions in Iran. Energy, 36(1), 220-225.

49. Herbst, A., Toro, F., Reitze, F., & Jochem, E. (2012). Introduction to energy systems modelling. Swiss Journal of Economics and Statistics, 148, 111-135.

50. Holdren, J. P., Smith, K. R., Kjellstrom, T., Streets, D., Wang, X., & Fischer, S. (2000). Energy, the environment and health. New York: United Nations Development Programme.41, 3654-3656.

51. Hong, M., Chen, Z. G., Yang, L., Zou, Y. C., Dargusch, M. S., Wang, H., & Zou, J. (2018). Realizing zT of 2.3 in Ge1− x− ySbxInyTe via reducing the phase-transition temperature and introducing resonant energy doping. Advanced Materials, 30(11), 1705942.

52. Hu, R., Krylov, S. Y., & Frenken, J. W. (2020). On the origin of frictional energy dissipation. Tribology Letters, 68, 1-13.

53. Huntington, H. G. (2010). Structural change and US energy use: recent patterns. The Energy Journal, 31(3), 651-655.

54. Ispas, A., Untaru, E. N., & Candrea, A. N. (2019). Environmental management practices within agritourism boarding houses in Romania: A qualitative study among managers. Sustainability, 11(14), 3887.

55. Jakob, M., Haller, M., & Marschinski, R. (2012). Will history repeat itself? Economic convergence and convergence in energy use patterns. Energy Economics, 34(1), 95-104.

56. Jiang, H. Y., Zhong, W., Wu, X. L., Tang, N. J., Liu, W., & Du, Y. W. (2004). Direct and alternating current magnetic properties of FeNi particles coated with SiO2. Journal of alloys and compounds, 384(1-2), 264-267.

57. Kannan, N., & Vakeesan, D. (2016). Solar energy for future world:-A review. Renewable and Sustainable Energy Reviews, 62, 1092-1105.

58. Kato, T., Yamada, Y., Nishikawa, Y., Otomo, T., Sato, H., & Sato, S. (2021). Origins of peaks of graphitic and pyrrolic nitrogen in N1s X-ray photoelectron spectra of carbon materials: quaternary nitrogen, tertiary amine, or secondary amine?. Journal of Materials Science, 56, 15798-15811..

59. Kemp, I. C. (2012). Fundamentals of energy analysis of dryers. Modern drying technology, 4, 1-45.

60. Kokaze, N., Nagasaki, K., Gotō, M., Hashimoto, Y., Muller, A. C., &

Shimoda, M. (2020). Toward a Model for Marking up Non-SI Units and Measurements. Journal of the Text Encoding Initiative, 4(12), 124-126.

61. Lewis, N. S. (2015). Introduction: solar energy conversion. Chemical Reviews, 115(23), 12631-12632.
62. Li, X., Wu, Y., Zhang, S., Cai, B., Gu, Y., Song, J., & Zeng, H. (2016). Quantum Dots: CsPbX3 Quantum Dots for Lighting and Displays: Room-Temperature Synthesis, Photoluminescence Superiorities, Underlying Origins and White Light-Emitting Diodes (Adv. Funct. Mater. 15/2016). Advanced Functional Materials, 26(15), 2584-2584.
63. LináWang, Z. (2014). Triboelectric nanogenerators as new energy technology and self-powered sensors–Principles, problems and perspectives. Faraday discussions, 176, 447-458.
64. Lior, N. (2008). Energy resources and use: The present situation and possible paths to the future. Energy, 33(6), 842-857.
65. Liu, D., Li, C., Tan, X., Lu, X., & Malik, O. P. (2021). Damping characteristics analysis of hydropower units under full operating conditions and control parameters: Accurate quantitative evaluation based on refined models. Applied Energy, 292, 116881.
66. Lloyd, B., & Subbarao, S. (2009). Development challenges under the Clean Development Mechanism (CDM)—Can renewable energy initiatives be put in place before peak oil?. Energy Policy, 37(1), 237-245.
67. Longfei, L. U., Jianzhong, Q. I. N., & Baojian, S. H. E. N. (2018). The origin of biogenic silica in siliceous shale from WufengLongmaxi Formation in the Middle and Upper Yangtze region and its relationship with shale gas enrichment. Earth Science Frontiers, 25(4), 226.
68. Magnus, A. M., West, J. S., Scott, D. W., Maxson, C., & Miller, J. (2019). Quantifying Qualitative Scholarship: A Fool's Errand or a Methodological Necessity?. Journal of Criminal Justice Education, 30(2), 223-249.
69. Maio, G. R., Verplanken, B., Manstead, A. S., Stroebe, W., Abraham, C., Sheeran, P., & Conner, M. (2007). Social psychological factors in lifestyle change and their relevance to policy. Social Issues and Policy Review, 1(1), 99-137.
70. Malen, J. A., Yee, S. K., Majumdar, A., & Segalman, R. A. (2010). Fundamentals of energy transport, energy conversion, and thermal properties in organic–inorganic heterojunctions. Chemical Physics Letters, 491(4-6), 109-122.
71. Mansoorian, H. J., Mahvi, A. H., & Jafari, A. J. (2014). Removal of lead and zinc from battery industry wastewater using electrocoagulation process: Influence of direct and alternating current by using iron and stainless steel

rod electrodes. Separation and Purification Technology, 135, 165-175.

72. Méplan, O., & Gignoux, C. (1996). Exponential growth of the energy of a wave in a 1D vibrating cavity: application to the quantum vacuum. Physical Review Letters, 76(3), 408.

73. Nakata, T. (2004). Energy-economic models and the environment. Progress in energy and combustion science, 30(4), 417-475.

74. Nordhaus, W. D., Houthakker, H., & Solow, R. (1973). The allocation of energy resources. Brookings papers on economic activity, 1973(3), 529-576.

75. Omer, A. M. (2008). Energy, environment and sustainable development. Renewable and Sustainable Energy Reviews, 12(9), 2265-2300.

76. Onile, A. E., Machlev, R., Petlenkov, E., Levron, Y., & Belikov, J. (2021). Uses of the digital twins concept for energy services, intelligent recommendation systems, and demand side management: A review. Energy Reports, 7, 997-1015.

77. Pang, X., Zhao, L., Feng, L., Meng, Q., Tang, X., & Li, J. (2009). The evolution and present status of the study on peak oil in China. Petroleum Science, 6, 217-224.

78. Panwar, N. L., Kaushik, S. C., & Kothari, S. (2011). Role of renewable energy sources in environmental protection: A review. Renewable and Sustainable Energy Reviews, 15(3), 1513-1524.

79. Patzek, T. W. (2008). Exponential growth, energetic Hubbert cycles, and the advancement of technology. Archives of Mining Sciences, 53(2), 131-159.

80. Pech, D., Brunet, M., Durou, H., Huang, P., Mochalin, V., Gogotsi, Y., ... & Simon, P. (2010). Ultrahigh-power micrometre-sized supercapacitors based on onion-like carbon. Nature Nanotechnology, 5(9), 651-654.

81. Perifanis, T. (2022). How US Suppliers Alter Their Extraction Rates and What This Means for Peak Oil Theory. Energies, 15(3), 821.

82. Poortinga, W., Steg, L., & Vlek, C. (2004). Values, environmental concern, and environmental behavior: A study into household energy use. Environment and behavior, 36(1), 70-93.

83. Purcell, A. H., & Smith, F. L. (1976). Energy and environmental impacts of materials alternatives: An assessment of quantitative understanding. Resource Recovery and Conservation, 2(2), 93-102.

84. Royapoor, M., Pazhoohesh, M., Davison, P. J., Patsios, C., & Walker, S. (2020). Building as a virtual power plant, magnitude and persistence

of deferrable loads and human comfort implications. Energy and Buildings, 213, 109794.

85. Sadler, B. M. (2005). Fundamentals of energy-constrained sensor network systems. IEEE Aerospace and Electronic Systems Magazine, 20(8), 17-35.
86. Schumack, M. (2006, June). The Hubbert Curve: Enabling Students to Meaningfully Model Energy Resource Depletion. In 2006 Annual Conference & Exposition (Vol. 2, pp. 11-1297).
87. Shabgard, H., Bergman, T. L., Sharifi, N., & Faghri, A. (2010). High temperature latent heat thermal energy storage using heat pipes. International Journal of Heat and Mass Transfer, 53(15-16), 2979-2988.
88. Shah, K., Turaev, D., & Rom-Kedar, V. (2010). Exponential energy growth in a Fermi accelerator. Physical Review E, 81(5), 056205.
89. Singh, H., Mishra, D., & Nahar, N. M. (2002). Energy use pattern in production agriculture of a typical village in arid zone, India—part I. Energy conversion and management, 43(16), 2275-2286.
90. Singh, H., Mishra, D., Nahar, N. M., & Ranjan, M. (2003). Energy use pattern in production agriculture of a typical village in arid zone India: part II. Energy Conversion and Management, 44(7), 1053-1067.
91. Soares, N., Costa, J. J., Gaspar, A. R., & Santos, P. (2013). Review of passive PCM latent heat thermal energy storage systems towards buildings' energy efficiency. Energy and buildings, 59, 82-103.
92. Son, Y. W., Cohen, M. L., & Louie, S. G. (2006). Energy gaps in graphene nanoribbons. Physical review letters, 97(21), 216803.
93. Sprehn, K. A., Whelton, M. G., Riley, D. R., & Okudan-Kremer, G. E. (2015). Content analysis of home energy audit reports: Quantitative methodology for standardization. Journal of Architectural Engineering, 21(4), B4015002.
94. Stirling, A. (2014). Transforming power: Social science and the politics of energy choices. Energy research & social science, 1, 83-95.
95. Talkington, S., Grijalva, S., & Reno, M. J. (2021). Power factor estimation of distributed energy resources using voltage magnitude measurements. Journal of Modern Power Systems and Clean Energy, 9(4), 859-869.
96. Tso, G. K., & Yau, K. K. (2003). A study of domestic energy usage patterns in Hong Kong. Energy, 28(15), 1671-1682.
97. Velloso, M., & Jardim, J. R. (2006). Study of energy expenditure during

activities of daily living using and not using body position recommended by energy conservation techniques in patients with COPD. Chest, 130(1), 126-132.

98. Vogelesang, H. (2008). An introduction to energy consumption in pumps. World pumps, 2008(496), 28-31.
99. Wang, C., Corbett, J. J., & Firestone, J. (2007). Modeling energy use and emissions from North American shipping: application of the ship traffic, energy, and environment model. Environmental science & technology, 41(9), 3226-3232.
100. Xu, G., Luo, L., Liang, J., Zhao, S., Yang, R., Wang, C., ... & Sun, X. (2022). Origin of high electrochemical stability of multi-metal chloride solid electrolytes for high energy all-solid-state lithium-ion batteries. Nano Energy, 92, 106674.
101. Yang, J., Ning, C., Deb, C., Zhang, F., Cheong, D., Lee, S. E., ... & Tham, K. W. (2017). k-Shape clustering algorithm for building energy usage patterns analysis and forecasting model accuracy improvement. Energy and Buildings, 146, 27-37.
102. Yin, W. J., Shi, T., & Yan, Y. (2014). Unique properties of halide perovskites as possible origins of the superior solar cell performance. Advanced materials, 26(27), 4653-4658.
103. Young, G. R. (1991). Energy conservation, occupational therapy, and the treatment of post-polio sequelae. Orthopedics, 14(11), 1233-1239.
104. Zaghi, S., Acar, M., Hultgren, B., Boggio, P. S., & Fregni, F. (2010). Noninvasive brain stimulation with low-intensity electrical currents: putative mechanisms of action for direct and alternating current stimulation. The Neuroscientist, 16(3), 285-307.
105. Zhong, H., Guo, M., Wang, Y., & Wang, Z. (2023). Quantify the magnitude and energy impact of overcooling in a sub-tropical campus building. Building and Environment, 231, 110033.
106. Zhu, J. Y., & Zhuang, X. S. (2012). Conceptual net energy output for biofuel production from lignocellulosic biomass through biorefining. Progress in Energy and Combustion Science, 38(4), 583-598.

CHAPTER

# 2 Forms of Energy

## Contents

## INTRODUCTION

The Sun's presence is essential for sustaining life on Earth. However, the likelihood of a planet orbiting a star possessing a biosphere, a narrow region capable of supporting life and facilitating its evolutionary processes, is exceedingly low. The Mars probes yielded no indications of life, while Venus, our neighboring planet, is inhospitably hot. The other planets in our solar system are even less conducive to supporting the only known form of life, which consists of carbon-based organisms. These organisms possess complex programs encoded in nucleic acids that enable their reproduction and survival, and they rely on enzymes to facilitate their metabolic processes. Despite the successful identification of numerous exoplanets, there is currently no evidence to suggest the presence of life on any of these celestial bodies. Furthermore, despite substantial investments in detecting extraterrestrial signals, our efforts have thus far yielded no communication from intelligent beings. Instead, the radio waves detected primarily originate from the ionized interstellar gas enveloping hot stars (Falk et al., 1983).

These findings are not particularly unexpected when analyzed solely from an energetic perspective. What is required is not merely a star (considering the vast number of approximately one hundred billion in our galaxy alone) accompanied by a substantial number of orbiting planets, but rather a star that meets the criteria of being appropriately sized, neither excessively large nor small, and possessing a temperature that falls within a favorable range, neither excessively cold nor hot. Stars that possess excessive mass have limited lifespans, which prevents the existence of complex life forms over billions of years, as required for the development of life on Earth. On the other hand, long-lived dwarf stars lack the necessary luminosity to provide sufficient energy for any potential planets in their orbit. A Goldilocks star must subsequently acquire a planet situated at an optimal distance that prevents the perpetual freezing of water (similar to Mars) or its vaporization (similar to Venus). This represents only the initial stage of an extensive catalog of necessary conditions that must be fulfilled to attain the circumstances conducive to supporting habitable planets or even the most rudimentary forms of life (McKendry, P. (2002).

**Did You Know?**

The understanding and harnessing of electricity as a form of energy started in the late 18th and early 19th centuries.

One effective approach to illustrate this concept involves contemplating the alteration of a single attribute that impacts the transmission of solar energy to Earth while assuming all other factors remain constant. For instance, consider the hypothetical scenario where the gravitational force exerted on Earth is doubled, or its orbital path deviates significantly from its current nearly circular trajectory. In the hypothetical scenario where the axis of rotation of the object in question was not inclined, the rotational period of 240 days, as opposed to the conventional twenty-four hours, and if the land area constituted ninety percent instead of thirty percent of its surface, what would be the implications? What would be the hypothetical scenario if the atmosphere of a given system lacked water vapor entirely and contained only a minimal concentration of carbon dioxide, specifically 0.038% (Kaper & Goedhart, 2002)?

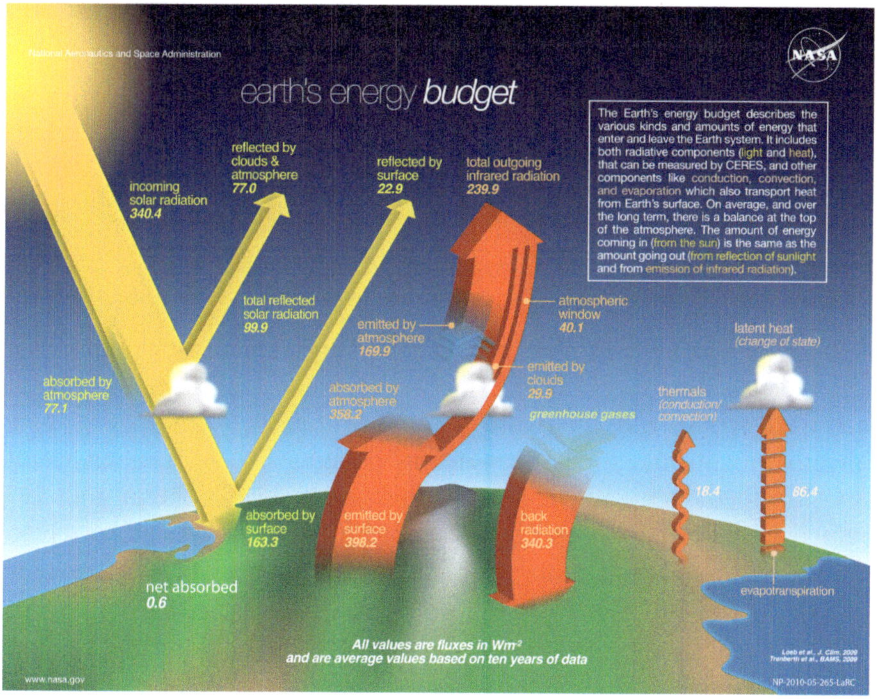


**Figure 2.1.** Representation of the earth's radiation balance (Source: By NASA - https://web.archive.org/web/20140421050855/http://science-edu.larc.nasa.gov/energy_budget/ quoting Loeb et al., J. Clim 2009 & Trenberth et al, BAMS 2009, Public Domain, https://commons.wikimedia.org/w/index.php?curid=32285340, Creative Commons License).

Without this absorption process, the Earth would exhibit characteristics of a "perfect black body radiator," resulting in re-radiating all the solar energy it intercepts. Consequently, the planet would maintain a temperature of 255 K (–18 °C). At a temperature of -18 °C, water would be solid, rendering it permanently frozen. Consequently, the conditions would be sterile to sustain any form of life (Bailey & McInnis, 2011).

The average radiative temperature of a planet is altered by the presence of atmospheric gases, which exhibit selective absorption of outgoing radiation and subsequently re-emit it in both upward and downward directions. This phenomenon contributes to the excessive heat experienced on Venus and the extreme cold on Mars. The "greenhouse effect" on Earth has played a crucial role in facilitating the evolution and diversification of complex life forms. This effect effectively increases the average surface temperature by 33 K, resulting in a mean temperature of 288 K (15 °C). This discussion pertains to the temperature experienced during a favorable spring day in a region with moderate climatic conditions. Additionally, it acknowledges the extensive presence of water, which covers over two-thirds of the Earth's surface, permeates soils and air, and constitutes a significant portion of the fresh weight of living organisms. Notably, water content can vary, with green plant tissues containing up to ninety-five percent and phytoplankton comprising approximately ninety-nine percent water (Bessman & Geiger, 1981).

After analyzing the trajectory of solar energy upon reaching the Earth, including the distribution of incoming radiation, its subsequent reflection back into space, and its transformation within the atmosphere, which collectively influence the Earth's climate patterns and fluctuations, we will now shift focus to the planet's sole significant non-solar energy source: its internal heat. Geothermal energy is responsible for driving the global tectonic processes, which result in the continuous reshaping of Earth's continents and oceans and the occurrence of highly destructive natural events such as volcanic eruptions, earthquakes, and tsunamis (Boyer, 2011).

Despite the remarkable diversity of living organisms in terms of their external characteristics, there is a limited set of fundamental metabolic pathways that these organisms utilize to convert energy from their surroundings into new living matter (Khan et al., 2019).

Autotrophs, referred to as primary producers, encompass all organisms capable of synthesizing fresh biomass from basic inorganic compounds. These autotrophs employ two distinct mechanisms to generate new biomass.

Phototrophs, including terrestrial plant and algae plankton, cyanobacteria, and purple and green sulfur bacteria, utilize electromagnetic energy to generate high-energy bonds of phosphate within ATP (adenosine triphosphate). ATP is the primary molecule responsible for storing and transporting energy within cells (Walker, 2008).

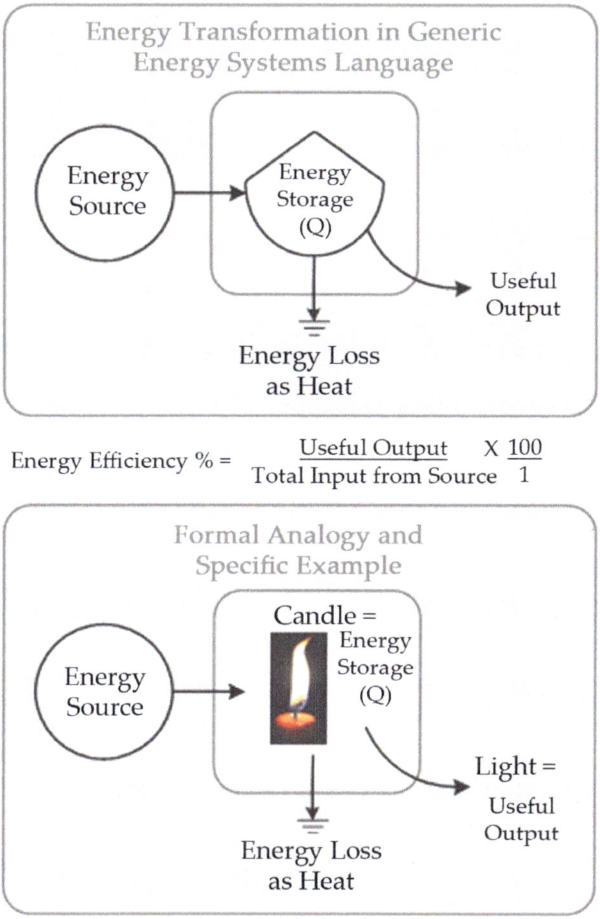


**Figure 2.2.** Illustration of the different forms of energy (Source: BY MR SHOLTO MAUD - OWN WORK, PUBLIC DOMAIN, HTTPS://COMMONS.WIKIMEDIA.ORG/W/INDEX.PHP?CURID=83697835, Creative Commons License).

These organisms harness this energy to facilitate the synthesis of new biomass (phyto mass) by assimilating atmospheric $CO_2$ and extracting essential macronutrients (such as nitrogen, phosphorus, and potassium) or micronutrients (such as iron, calcium, silicon, and others) from the soil. Chemotrophs, such as nitrifying bacteria responsible for converting ammonia to nitrates, iron, non-pigmented sulfur, and methane-producing microbes, do not rely on light for their energy needs. Instead, they require carbon dioxide, oxygen, and either an oxidizable element like hydrogen or iron or a simple inorganic compound such as hydrogen sulfide or ammonia. The imperceptible metabolic processes of organisms play a vital role in the biosphere's essential biological, geological, and biogeochemical cycles (Watts, 1983).

Heterotrophs, also known as chemoorganotrophs, are organisms that cannot produce new biomass from basic inorganic substances and instead rely on the assimilation of digested organic compounds for their essential building blocks (Carminati et al., 1993).

## 2.1 SOURCES OF ENERGY IN NATURE

They are detailed in the following section:

### 2.1.1 Sun and Earth: Solar Radiation and its Return

Astronomers often emphasize that the Sun is classified as a G2 dwarf star, considered a prevalent type of star. These stars are not particularly notable in size or radiation output. Their primary energy source stems from the proton-proton fusion reaction, wherein hydrogen atoms combine to form helium under temperatures exceeding 13 million Kelvin. The Sun's energy production, also known as its total luminosity, is significant compared to terrestrial standards. This is due to thermonuclear reactions within its core, which convert approximately 4.4 million tons of matter into energy per second. By applying Einstein's mass-energy equation, this calculation results in an output of nearly 3.9 × 10^26 watts, a rate that surpasses our utilization of all fuel sources (including fossil and biomass) as well as primary electricity (hydro and nuclear) in the year 2005 by approximately 30 trillion times, or thirteen orders of magnitude (Gallo et al., 2012).

Approximately 4.5 billion years ago, during the formation of the Earth, the luminosity of the nascent Sun was approximately 30% lower than its present-day value. During this particular period, the sun has only consumed a mere 0.03% of its vast mass, yet it has managed to deplete over half of the hydrogen present in its core. The future trajectory of the sun holds minimal significance for our civilization, as it is projected to cease to exist well before undergoing its transformative stages. These stages involve the sun expanding into a red giant, reaching a diameter approximately one hundred times its current size, which will engulf the planet with the sun's intense energy. Subsequently, the sun will contract into a highly luminous white dwarf. It is important to note that the sun's evolutionary epochs are measured in billions of years. In contrast, the existence of complex civilizations has thus far been limited to approximately 5,000 years (Chen et al., 2013).

Without clouds, Earth's average annual solar radiation would exhibit a consistent decrease towards the poles. However, tropical clouds significantly reduce the amount of solar energy reaching the equatorial region, while monsoonal clouds have a similar effect on the northern regions of Asia (Lewis et al., 2005).

As a result, significant portions of equatorial Amazonia, southern Nigeria (located just 5° north of the equator), and provinces in the southern half of China (particularly the landlocked Sichuan, situated in a mountain basin at 30° N) experience lower annual sunlight exposure compared to New England, a region spanning from 40 to 45° N. Furthermore, it is noteworthy that the zenith points of midday summer solar radiation are remarkably similar in Jakarta, the capital city of Indonesia, situated at a latitude of 6° S, and Edmonton, the capital city of Alberta, located at a latitude of approximately 55° N. The realities above have significant implications for future endeavors involving the large-scale direct conversion of insolation into electricity through photovoltaic means (Dupont et al., 2020).

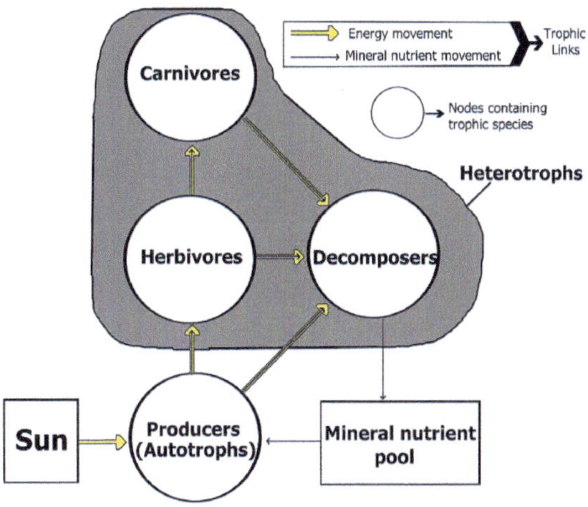


**Figure 2.3.** Solar energy utilization (Source: By Thompsma - Own work, CC BY-SA 3.0, https://commons.wikimedia.org/w/index.php?curid=15521895, Creative Commons License).

### 2.1.2 Air and Water: Avenues in Motion

Radiation absorption plays a crucial role in providing three essential energy services. Firstly, it contributes to the heating of continents and oceans. The subsequent re-radiation of this heat is primarily responsible for warming the atmosphere, thereby maintaining its continuous motion. Secondly, absorbed radiation facilitates the evaporation of water, enabling its distribution over long distances from its sources. Lastly, absorbed radiation energizes the process of photosynthesis. Due to the comparably low specific mass of air, with a mass of approximately 1.2 kg per cubic meter near the Earth's surface, which is one-thousandth of the mass of water, a minimal proportion of incoming solar radiation, potentially as little as two percent, suffices to fuel the worldwide atmospheric circulation. This circulation plays a crucial role in heat dispersion, transportation of microorganisms, pollen, and seeds, and the wind-induced erosion of continental landforms. The global atmospheric circulation is driven by the persistent heating of the tropics, resulting in the movement of cooler air from higher latitudes toward the equator. This phenomenon, known as the intertropical convergence zone, initiates the formation of two dynamic air circulation patterns referred to as Hadley cells. These cells were initially identified by George Hadley, an English physicist, who provided the first description of their presence (Occhiello et al., 1991).

**Did You Know?**

The first wind turbine built for electricity generation was constructed in Cleveland, Ohio, by engineer Charles Bush in 1887.

The high velocities of near-ground winds result from the significant thermal energy accumulation during the summer season, leading to cyclonic atmospheric circulations characterized by low pressure. These circulations can range from small, harmless thunderstorms to large-scale hurricanes. Typically, even thunderstorms of considerable magnitude, generating power ranging from ten to a few hundred gigawatts, do not generate winds that exert vertical power densities exceeding 15 kW/m^2. It is worth noting that this threshold remains below the level at which structural damage is typically observed. North American hurricanes have their genesis in Africa, initially proceeding in a westerly direction before deviating clockwise, often resulting in their eventual landfall across the northern Gulf of Mexico, Florida, and the East Coast. Typhoons, the Asian counterparts of hurricanes, typically form over the warm waters of the Pacific Ocean near the Mariannas. These weather systems then proceed westward, frequently impacting extensive regions of Southeast Asia, the coastline of China, the Korean peninsula, and Japan. Hurricanes or cyclones possess velocities reaching up to 90 m/s (equivalent to over 300 km/h) and impact vertical surfaces with power densities of up to 1 MW/m^2. Contemporary steel and concrete structures can effectively withstand these forces but pose a challenge for wooden-framed houses (Mahto et al., 2021).

While certain hurricanes can persist for extended periods, impacting consecutive large regions along a path spanning several thousand kilometers, tornadoes are comparatively more localized in their scope. The typical trajectory of a tornado in the United States is relatively narrow, measuring approximately 125 meters in width. Often, there is a distinct boundary between areas of minimal damage, where houses may remain largely intact, and areas of destruction, where structures are severely impacted. Furthermore, the length of these tornado paths is typically less than 10 kilometers, and their duration is generally under three minutes. On the contrary, tornadoes from the most severe (and comparatively infrequent) classification can produce wind speeds surpassing 100 m/s and impact vertical structures with greater force than a typical hurricane (Silverstein & Fort, 2000).

Merely a fraction of continental precipitation replenishes deep aquifers, with approximately 60% subject to evaporation and less than 33% discharged into the ocean through streams. With an average continental elevation of 850 m, this stream flow possesses an annual potential gravitational energy of approximately 400 EJ (13 TW). This value is ten times greater than the global electricity consumption at the onset of the twenty-first century. The utilization of the vast potential of hydroelectricity generation is constrained due to various factors. These include the limited availability of suitable sites for constructing large dams, competing demands for water, such as irrigation, urban areas, and industrial usage, and the need to maintain minimum stream flows to sustain aquatic ecosystems and remove sediment (Prommas et al., 2010).

### 2.1.3 The Earth's Heat

The alternative energy source that powers our planet, although significantly smaller in magnitude compared to solar radiation, has had an immeasurable qualitative influence on the development of life and the outcomes of civilizations. This is due to the continuous process of the Earth's internal heat, which constantly generates new oceanic crust and rearranges and separates continents. Recurring and devastating events, such as intense seismic activity, transoceanic tsunami occurrences, and remarkable volcanic eruptions, frequently accompany these significant geotectonic processes. These events substantially alter landforms and result in the loss of human lives. The internal energy of the Earth can be attributed to two primary sources. Firstly, it arises from the basal heat from the gradual cooling of the Earth's molten metallic core, predominantly composed of iron. Secondly, it is derived from radioactive decay, with notable contributions from uranium isotopes 235 and 238, thorium 232, and potassium 40. The latter flux holds greater significance, and although a precise delineation of the sources of heat remains elusive, an ample array of measurements exists to support the assertion that the cumulative worldwide capacity of geothermal energy reaches approximately 44 terawatts (TW) (Spedding & Henderson, 1971).

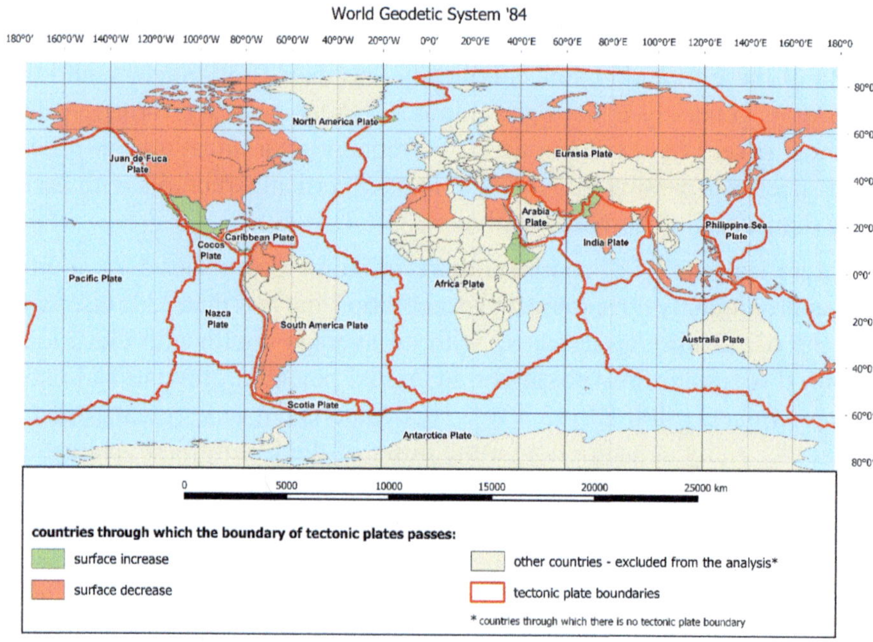


**Figure 2.4.** Illustration of the geotectonic plates (Source: Maciuk K, Apollo M, Kukulska-Kozieł A, Lewińska P. What Is the Impact of Tectonic Plate Movement on Country Size? A Long-Term Forecast. Remote Sensing. 2021; 13(23):4872. https://doi.org/10.3390/rs13234872, Creative Commons License).

Approximately 60% of the Earth's thermal energy creates fresh oceanic crust spanning approximately 55,000 km along the ocean ridges. These ridges serve as boundaries between the Earth's crust, a thin and solid outer layer, and the geotectonic plates, which are rigid and undergo gradual movement

(Figure 2.2). The Pacific plate, which is the largest among its counterparts, consists entirely of oceanic material and exhibits a relatively thin structure, measuring less than 10 km in certain regions. In contrast, other tectonic plates possess continental masses attached to them, resulting in crustal thicknesses exceeding 100 km. Basaltic magma, which ascends from the underlying mantle along the ridges, generates approximately three square kilometers of fresh oceanic crust at an average global spreading rate of less than 5 cm per year. The manifestation of the rifting process can be observed on land in several locations worldwide, including the Afar region between Eritrea and Somalia, the East African Rift Valley, and central Iceland. Oceanic plates diverging necessitates their eventual collision with larger continental plates, resulting in the recycling (subduction) of the ocean floor back into the Earth's mantle. The mantle, a solid layer approximately 3,000 km thick, is between the crust and the liquid core. The most remarkable characteristics of plate subduction are observed in deep ocean trenches (Stein et al., 2013).

The absence of oceanic crust exceeding approximately 200 million years in age, with the majority being younger than 100 million years, can be attributed to the ongoing recycling process. This phenomenon also accounts for the concentration of highly intense earthquakes, frequently resulting in significant tsunami events and volcanic eruptions along the subduction fronts. The Pacific plate, which moves relatively quickly, is subducted beneath the relatively stationary Australian and Eurasian plates. This process gives rise to a vast semi-circular arrangement of deep ocean trenches from the Aleutian Islands to Tonga, located north of New Zealand (Jennings et al., 1960).

Due to their sporadic and brief nature, volcanic eruptions contribute only a minor portion of the overall global output of geothermal energy. Current estimations suggest that volcanic eruptions account for approximately two percent of the total geothermal energy flux. However, significant variability exists yearly, as there can be decades-long intervals between large-scale volcanic events. Moreover, the duration and intensity of volcanic eruptions can vary greatly, with some volcanoes exhibiting violent but short-lived eruptions while others remain active for extended periods (Moeller & Ferrus, 1961).

Heat typically significantly influences the overall discharge of volcanic energy, surpassing it by a factor of one to three orders of magnitude. However, the primary conveyors of this heat differ. In numerous volcanic occurrences, the predominant mechanism for heat transfer involves the ascent of substantial ash clouds, which can reach the stratosphere. This phenomenon was notably observed during the eruption of Mount Pinatubo in the Philippines on June 15, 1991. The resulting ash clouds induced a prolonged global decrease in surface temperatures for several months (Valencia & O'Connell, 2009).

On the other hand, Hawaiian volcanoes expel their thermal energy through the emission of sluggish lavas, characterized by their smooth, rope-like appearance known as pahoehoe. Certain flows of these lavas allow for proximity to facilitate the collection of samples from the high-temperature magma. The pyroclastic flows, which involve the amalgamation of volcanic substances spanning from fine ash to sizable rocks, along with exceedingly hot gases surpassing temperatures of 500 °C, are unequivocally the most dangerous manifestations of heat release. The velocity at which they descend downhill exceeds 100 km/h, resulting in the potential to engulf and cover objects within a radius of up to 100 km. The eruption of Mount Pelée in 1902 resulted in the demise of 28,000 individuals residing in St. Pierre, Martinique, due to the occurrence of nuées ardentes, also known as glowing clouds. In August 1997, extensive devastation occurred on the Caribbean island of Montserrat due to the destructive impact of pyroclastic flows. Similarly, recurrent instances of pyroclastic flows have been documented at Unzen in Japan (Jaupart et al., 2007).

Similar to seismic activities, volcanoes are predominantly linked to the boundaries of tectonic plates. However, in certain instances, intense molten magma plumes have managed to breach a plate, resulting in remarkable hot spots that are geographically distant from any subduction or collision regions. One prominent illustration is the series of Hawaiian Islands, which stretch as seamounts until reaching Kamchatka. The formation of this volcanic chain is attributed to a substantial hot spot, which consistently intersects the Pacific plate as it moves in a northwestern direction. The hot spot is located beneath the western coast of Hawaii, resulting in ongoing volcanic activity at the Kilauea volcano (Van Heck & Tackley, 2011).

Additionally, the hot spot is positioned just offshore from the island where Loihi, a significant submarine volcano, will eventually emerge. However, it is estimated that this emergence will not occur for tens of thousands of years, making Loihi the most recent addition to the volcanic chain. Africa's plate experiences another significant tectonic activity in the central region of the continent, resulting in the formation of the Virunga volcano chain. This volcanic chain is located along the borders of Uganda, Rwanda, and Zaire.

Notably, the bamboo forest in the foothills of Mount Karisimbi, one of these volcanoes measuring approximately 4,500 meters in height, is the habitat for the world's remaining mountain gorillas (Gupta & Krishnamurthy, 1992).

### 2.1.4 Photosynthesis: Reactions and Rates

The process of photosynthesis is initiated through the uptake of light by pigments located in the membranes of the thylakoid within the chloroplasts of bacteria and plants, which are the cellular organelles responsible for the green coloration observed in plants. The energy efficiency of converting basic inorganic inputs into fresh phyto mass exhibits a remarkably low level. In introductory textbooks, it is common to find a concise representation of the overall process through a simplified equation. This equation illustrates the reaction involving six molecules of carbon dioxide (CO2) and six molecules of water (H2O), resulting in the production of one molecule of sugar and six molecules of oxygen (O2). The chemical equation presented is a representation of the process of photosynthesis, where carbon dioxide (CO2) and water (H2O) react to produce glucose (C6H12O6) and oxygen (O2) (Wasielewski, 1992).

$$6CO_2 + 6H_2O \rightarrow C_6H_{12}O_6 + 6O_2$$

Chlorophylls a and b, the primary pigments in plants capable of being stimulated by radiation, exhibit relatively limited absorption peaks. Specifically, the absorption maximum for chlorophyll a falls within the range of 420–450 nm, while that of chlorophyll b falls within the range of 630–690 nm. A synergistic interaction between blue and red light primarily fuels photosynthesis. Consequently, the pigments involved in photosynthesis exhibit minimal absorption of light in the green and yellow regions of the visible spectrum. As a result, during spring and summer seasons, the reflected green and yellow colors from leaves become prominent, remaining relatively stable until the pigments decompose in the autumn (Emerson & Arnold, 1932).

Additionally, it should be noted that photosynthetically active radiation (PAR) constitutes approximately forty-three percent of the total insolation. The pigments absorb energy, which subsequently facilitates the electron transport process. Water acts as the electron donor, thereby serving as the origin of oxygen. This electron transport involves three multi-enzyme complexes and synthesizes NADP (nicotinamide adenine dinucleotide phosphate) and ATP (adenosine triphosphate). NADP and ATP play crucial roles as enzymes within cells. They drive the assimilation of carbon derived from CO2 into carbohydrates. This assimilation process follows three distinct pathways (Fleming et al., 1988).

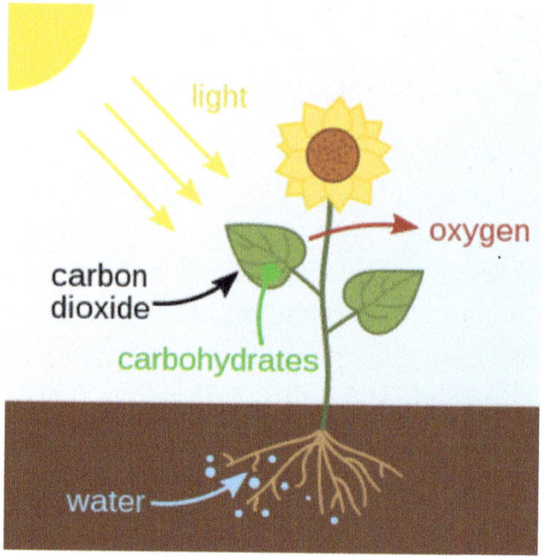


**Figure 2.5.** Illustration of the process of photosynthesis (Source: Https://en.wikipedia.org/wiki/File:Photosynthesis_en.svg, Creative Commons License).

Irrespective of the pathway, it will inevitably allocate a substantial portion of its fixed carbon toward respiration. This metabolic process generates energy through the oxidation of sugars. The process above occurs within specialized organelles known as mitochondria, located within plant cells, wherein the derived energy is utilized to sustain fundamental cellular activities. The processes encompassed within this category involve the transportation of sugars or starches, also known as photosynthate, from leaves to stems and roots. Additionally, it encompasses the absorption of macronutrients and micro-nutrients from the soil and their incorporation into organic compounds. Respiration serves as a means of acquiring energy, which is subsequently utilized in synthesizing intricate organic compounds essential for the metabolic processes of organisms (Maroti & Wraight, 1997).

The maximum theoretical net efficiency, accounting for all respiration losses, is approximately four percent of insolation. However, this efficiency level can only be achieved during short intervals and under favorable conditions of sufficient water and nutrients. Cultivated crops that receive intensive care, such as irrigation and fertilization, typically exhibit an average of two percent efficiency throughout their growth period. In comparison, the most profitable temperate and tropical forests achieve an efficiency of approximately 1.5 percent. The average global continental value stands at a mere 0.33% (Baly, 1935).

Furthermore, the conversion of insolation into new aquatic phyto mass by oceanic phytoplankton is less than 0.1%. Consequently, the overall average for the biosphere is below 0.2%. In other words, a mere 0.2% of the photons that reach the planet's ice-free surface are transformed into new phyto mass. The significant influx of solar radiation renders this inefficiency largely inconsequential. Photosynthesis's overall efficiency is noteworthy in terms of quantity and quality differentiation. The most accurate global estimations indicate an annual net primary production (NPP) of approximately 120 billion tonnes on landmasses and 110 billion tonnes in marine environments (Bulthuis, 1987).

Forests are widely recognized as being highly productive ecosystems. The spatial arrangement of major biomes, which are intricate and extensive communities of organisms, is primarily constrained by temperature and precipitation rather than insolation. For instance, tropical rainforests necessitate a minimum annual rainfall of one meter and an average yearly temperature exceeding 20 °C. On the other hand, summer green (deciduous) forests, which prevail as the primary natural vegetation in Western Europe and Eastern North America, occupy a wide ecological niche characterized by annual average temperatures ranging from 0 to 20 °C and rainfall varying between less than half a meter and over two meters per year. The annual Net Primary Productivity (NPP) rates, measured in terms of absolute dry matter, exhibit significant variation across major ecosystems. For instance, tropical rainforests demonstrate a wide range of 1-3.5 $kg/m^2$ (equivalent to 10-30 t/ha) in NPP, while temperate forests exhibit 0.5-2.5 $kg/m^2$ rates. Similarly, most grasslands display NPP rates of 0.2-1.5 $kg/m^2$ (Myers, 1971).

### 2.1.5 Heterotrophs: Metabolism and Locomotion

Heterotrophs possess two fundamental mechanisms for metabolizing complex organic compounds: anaerobic and aerobic. The two categories of anaerobic fermenters that hold significant relevance in human affairs encompass the bacterial agents accountable for lactic fermentation and yeasts. Bacterial fermentation, a process in which sugars are converted into lactic acid, yields various food products such as sour milk, yogurt, cheese, sauerkraut, gherkins, and olives. Yeasts, classified as fungi within the phylum Ascomycota, can ferment alcohol using a diverse range of sweet or starchy substances as substrates. The anaerobic pathway emerged in the early stages of Earth's history, characterized by an atmosphere containing minimal oxygen levels (Evans et al., 2009).

In contrast, the aerobic pathway became prevalent following the development of an oxygen-rich atmosphere over the last two billion years. When oxygen is available, yeasts undergo a metabolic process that produces carbon dioxide rather than alcohol. This carbon dioxide production facilitates dough leavening, leading to leavened bread instead of unleavened bread varieties such as pita or chapati. Approximately 50% of bacterial phyla exhibit aerobic characteristics. This group encompasses economically significant species, particularly the nitrogen-fixing bacteria, predominantly those belonging to the genus Rhizobium. These bacteria play a crucial role in providing leguminous plants with the essential nitrogen nutrient of ammonia while receiving a carbohydrate supply. All organisms classified under the taxonomic kingdom Animalia are characterized by their ability to obtain energy through aerobic respiration and rely on external sources for organic compounds. The initial objective of heterotrophic metabolism involves the enzymatic degradation of carbohydrates into their monosaccharide units, namely glucose, and fructose. It is worth noting that sucrose, a prominent sweetening agent globally, is a disaccharide formed by the combination of these monosaccharides (Makarieva et al., 2008).

Additionally, this metabolic process entails the hydrolysis of lipids, wherein they are broken down into glycerol and fatty acids through a reaction facilitated by the presence of water. Lastly, proteins are decomposed into constituent amino acids, subsequently utilized to synthesize new protein structures such as muscles, organs, and cells. Adenosine triphosphate (ATP) serves as an energy reservoir by capturing the energy released during the breakdown of nutrients. Through its conversion to adenosine diphosphate (ADP), ATP provides the necessary energy to synthesize fresh biomass and facilitates locomotion (Crawford, 1992).

A multitude of enzymes facilitate the catalysis of these intricate reactions. During fermentation, the ultimate outcomes consist of either lactic acid or ethanol and CO2. In the case of champagne production, the CO2 gas is retained, whereas it is not preserved in the production of table wines. Conversely, aerobic metabolism yields substantial quantities of ATP and releases CO2 (exhaled) and water (eliminated through respiration, perspiration, urination, and defecation). The anaerobic fermentation process exhibits a maximum overall efficiency of approximately thirty percent, while aerobic metabolism demonstrates a higher efficiency of around sixty percent. This efficiency applies to the conversion of both simple sugars and fatty acids. Each heterotroph possesses a basal metabolic rate (BMR), which refers to the constant minimum energy requirement necessary to sustain the functioning of its essential internal organs (Raina et al., 2023).

Heterotrophs with limited mobility, such as many invertebrates and certain reptiles and fish, exhibit an overall energy expenditure that is only a moderate multiple of their basal metabolic rate (BMR). In contrast, highly active fliers and extensively roaming mammals allocate a significant portion of their energy toward fulfilling the demands of their strenuous activities. Extended observations reveal that rodents exhibit a daily metabolism approximately 2.5 times higher than their basal metabolic rate (BMR). Conversely, avian species that engage in aerial foraging display a total daily metabolism exceeding four times their BMR. The metabolic scope of heterotrophs imposes limitations on their running, flying, and swimming speeds and endurance. This scope refers to the ratio between their short-term maximum energy expenditure and basal metabolic rate (BMR). The average range of metabolic rates for mammals, regardless of their body mass, is approximately ten. In contrast, certain fish and most birds exhibit a range of about fifteen, although it should be noted that avian basal metabolic rate surpasses that of fish or mammals. Conversely, reptiles and amphibians typically demonstrate a range of less than five (Tan et al., 2019).

Unsurprisingly, when examining the energy expenditure associated with the three primary modes of locomotion, it becomes evident that swimming, owing to the buoyancy of the water, is the most energy-efficient means of movement. Swimming necessitates less than one-tenth of the energy required for running, whereas running is more energetically demanding than flying for organisms weighing up to 1 kg. In the case of avian species with greater body mass, the disparity diminishes, as the upper limit of the largest flying birds' mass is constrained by the phenomenon wherein the power required for flight escalates faster than the power that the pectoral muscles can generate. As a result, the ability of larger avian species to achieve flight is hindered, leading them to employ specific strategies such as initiating takeoff in the presence of strong winds from elevated locations such as cliffs or steep hillsides. Notable examples of this behavior can be observed in the American condor and royal albatross. A limited number of avian species possess a weight exceeding ten kilograms. Among these, the East African Kori bustard, weighing 13 to 19 kilograms, exhibits infrequent instances of achieving flight, primarily through a sequence of low-altitude hops (Fenchel & Finlay, 1983).

The facilitation of long-distance migrations through locomotive means such as running, swimming, and flying is significantly enhanced when animals can forage during their journey, as observed in caribou, salmon, and Canadian geese. However, paradoxically, it has been observed that songbirds undertaking migration in colder weather may expend more energy during their stopovers than during the actual flight itself. The most remarkable long-distance migration is observed in small songbirds, who undertake uninterrupted flights across vast stretches of the ocean during their journey from northern and temperate breeding grounds to subtropical and tropical wintering locations. The migrations observed in certain organisms, which are influenced by celestial bodies such as stars and the sun, as well as the Earth's geomagnetic field and polarized light, are hypothesized to be facilitated by utilizing fat as a highly efficient energy source for the body. Therefore, the potential range (excluding any favorable wind conditions) of a diminutive songbird's non-stop flight will be constrained by its capacity to accumulate fat before departure. Numerous species experience an increase of up to thirty percent in their typical body mass and exhibit a near absence of energy reserves upon arrival at their remote destination, often resulting in their exhaustion and subsequent collapse on beaches (Strumpen et al., 2022).

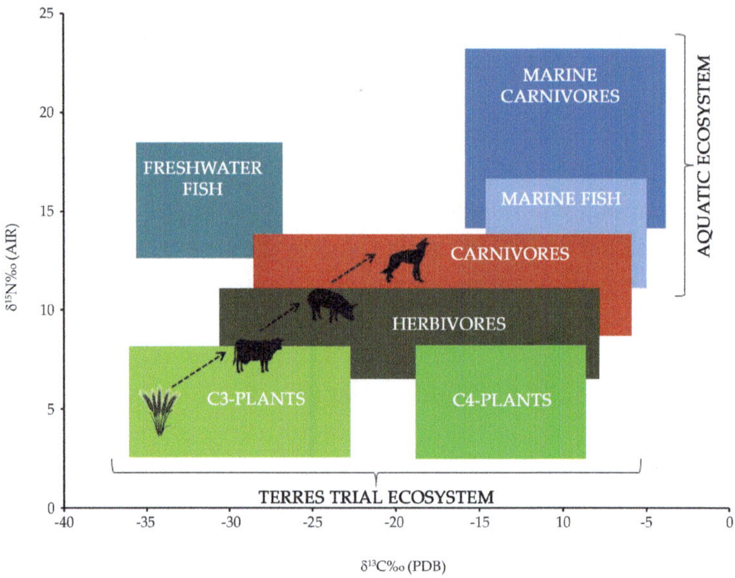


**Figure 2.6.** Representation of the types of heterotrophic nutrition (Source: Alt KW, Al-Ahmad A, Woelber JP. Nutrition and Health in Human Evolution–Past

to Present. Nutrients. 2022; 14(17):3594. https://doi.org/10.3390/nu14173594, Creative Commons License).

### 2.1.6 Energy in Ecosystems

Thermodynamic imperatives govern the energy that passes through ecosystems. Although photosynthesis is considered relatively inefficient, autotrophs occupy the highest position in the energy conversion hierarchy of the biosphere. Consequently, the overall biomass of an ecosystem is consistently dominated by the phyto mass of autotrophs. As the distance between a living organism and the primary solar energy source increases, the amount of energy accessible at each subsequent trophic level diminishes. It is important to note that the commonly used term "food chain" is often misleading, as ecological relationships are better represented as interconnected webs rather than linear chains (Baldocchi, 2014).

It is evident, without the requirement of scientific studies, that herbivores, which encompass primary consumers such as rodents and ungulates, exhibit significantly higher population densities compared to secondary consumers, including carnivores such as birds of prey that prey on rodents or ladybugs that consume aphids. In contrast, tertiary consumers are comparatively less abundant, including secondary carnivores like aquatic birds that prey on frogs consuming insects or predatory fish. Furthermore, it is important to note that the ability to consume a wide range of autotrophs, heterotrophs, and decomposing biomass greatly improves one's access to sustenance as an omnivorous organism. This dietary flexibility allows for consuming any available food source readily located, captured, and efficiently digested (Odum, 1968).

Omnivory is a prevalent phenomenon observed in various species, primarily driven by the availability of resources and the necessity to acquire them. This ecological behavior is frequently observed when herbivorous or carnivorous organisms opportunistically shift their trophic level, taking advantage of temporary abundance in certain resources or compensating for scarcity in others (Picchi et al., 2019).

Decomposers are integral components of ecosystems, as they facilitate the breakdown of complex organic macromolecules and repeatedly release vital macro and micro-nutrients, which are subsequently accessible to autotrophs and heterotrophs alike. It is worth noting that decomposers can consume deceased organisms across various trophic levels (Hastik et al., 2015).

In most terrestrial ecosystems, the trophic webs exhibit limited trophic levels. In the grasslands of East Africa, which are known for their high population of large animals, the majority of energy transfer is limited to two trophic levels. The primary consumers in this ecosystem, including ungulates such as antelopes, gazelles, impalas, and wildebeest, primarily consume their energy by consuming grass. The secondary consumers in this system, which consist of felids such as leopards, cheetahs, and lions, as well as canids like wild dogs and hyenas, derive their energy by preying on the ungulates. Finally, decomposers play a crucial role in this ecosystem by feeding on the remains of these organisms and any other organic matter that they can break down using their enzymes. No predatory species of significant aggression inhabit the sixth level of the solar system for sustenance. In tropical rainforests characterized by a higher and more diverse standing phyto mass, as well as a greater abundance of heterotrophs, it is observed that three trophic levels are commonly present. In comparison, five trophic levels are potentially achievable. Specifically, the trophic interactions involve fungi as primary consumers, arboreal invertebrates as secondary consumers, frogs as tertiary consumers, snakes as quaternary consumers, and birds as top consumers, preying on snakes (Hannon, 1973).

Marine ecosystems rely on primary production facilitated by phytoplankton, a diverse group of small autotrophic organisms encompassing bacteria, cyanobacteria, archaea (unicellular organisms that share similarities with bacteria but possess distinct genetic characteristics), and algae. The complexity of marine food webs tends to surpass that of terrestrial biomes. The number of trophic levels in certain ecosystems can reach up to five, and it can extend to six in kelp forests. However, the most intricate coral reefs have been observed to support up to seven trophic levels, showcasing unparalleled complexity (Porter et al., 2009).

A comprehensive depiction of biomass in a given unit area of a terrestrial ecosystem would exhibit a pyramid-shaped pattern, wherein autotrophs constitute a wide foundational base, followed by herbivores occupying a smaller second tier. Above them, omnivores and first-order carnivores occupy the subsequent level, while a limited number of top predators occupy the apex. The mass distribution across different ecosystems exhibits significant variation, with phytoplankton mass typically being twenty times greater than the mass of primary consumers. In contrast, the mass of top carnivores may constitute less than 0.001% of the phytoplankton mass (Ranius et al., 2018).

In marine ecosystems, an inverted pyramid structure is observed, wherein the relatively short lifespan of phytoplankton, typically ranging from 12 to 48 hours, coupled with the substantial consumption rates by zooplankton and larger herbivores, results in the total standing heterotrophic biomass potentially exceeding the mass of the photosynthesizing phytoplankton by a factor of two to four. The collective truth is that most autotrophic organisms in the ocean are microscopic phytoplankton with an average diameter of approximately 10 μm. Conversely, higher trophic levels consist of zooplankton as primary consumers, small fish as secondary consumers, larger fish and common squid as tertiary consumers, and tuna as quaternary consumers. It is worth noting that the size of these organisms increases progressively, with mature Southern Bluefin tuna, for instance, weighing over 150 kg before their population was significantly reduced due to overfishing. There exist noteworthy exceptions. The most sizable marine organisms, such as blue whales with a weight of up to 130 metric tons, and fish species like the whale shark, which can weigh up to 1.5 metric tons, exhibit filter-feeding behavior. These organisms consume substantial quantities of minute phytoplankton and zooplankton (Zeng et al., 2005).

The decrease in the population of heterotrophs in higher trophic levels is frequently linked to an escalation in body size. Typically, top predators encompass the largest creatures within their respective categories, such as golden eagles among avian predators or tigers and lions among felids. Herbivory confers evident energetic benefits, and within contemporary ecosystems, the organisms exhibiting the greatest physical size are classified as megaherbivores. These megaherbivores, characterized by a body mass exceeding one tonne, include elephants, hippos, giraffes inhabiting tropical regions and moose and muskoxen in boreal and Arctic environments. The dominance of megaherbivores, both extinct and extant, was particularly evident in earlier periods, wherein the largest individuals, such as the mammoth and colossal dinosaurian weighing over 80 metric tons, exhibited a significantly greater magnitude of mass compared to the heaviest species found in contemporary times (De Strasser et al., 2016).

## 2.2 KINETIC ENERGY

The concept of kinetic energy is a fundamental principle within the field of physics, which elucidates the energy transferred to an object as a direct consequence of its motion. To induce acceleration in an object, it is necessary to exert a force, which entails the execution of work. When external forces are applied to an object, work is performed, leading to the transfer of energy to the object, consequently causing a change in its velocity. The energy that is transferred is commonly referred to as kinetic energy, which is contingent upon both the mass of the object and the velocity it attains. Few examples of kinetic energy in action include an individual engaged in walking, a baseball soaring through the air, a crumb descending from a table, and a charged particle navigating through an electric field. An object in a state of rest possesses a kinetic energy value of zero (Wang & Teter, 1992).

As per the physics definition, the kinetic energy of an object is quantified as the capacity to perform work due to its motion. Notably, kinetic energy is a scalar quantity characterized solely by its magnitude and devoid of any directional aspect. In physics, it is widely acknowledged that stationary objects are endowed with potential energy, considered one of the primary forms of energy. The conversion of potential energy into kinetic energy occurs when an external force, such as gravity, is applied to an object, initiating motion. One example of potential energy is elastic potential energy, which is stored in a rubber band that has been stretched. The potential energy is transformed into kinetic energy when the rubber band is released (Fjørtoft, 1953).

As per the principle of energy conservation, it is postulated that energy can neither be generated nor annihilated. Energy can solely undergo conversion from one form to another. An illustration of this phenomenon can be observed in the oscillatory motion of a pendulum, wherein the pendulum alternates between possessing kinetic energy and potential energy. At the apex of the swing, when the pendulum experiences a momentary state of rest, its energy is in the form of potential energy, with no kinetic energy. At the bottom of the pendulum's oscillation, its velocity reaches its maximum, resulting in the entirety of its energy being kinetic, with no presence of potential energy. Throughout the entire oscillation of a pendulum, the total energy remains conserved as the conversion between potential energy and kinetic energy occurs cyclically (Khaligh et al., 2009).

In summary, kinetic energy (KE) refers to the energy linked to an object's movement, which is acquired through the application of work. The relationship between the force exerted on an object and its kinetic energy is contingent upon both the mass of the object and the square of its velocity. A comprehensive understanding of kinetic energy is essential for grasping a wide range of physical phenomena (Tal & Bader, 1978). The kinetic energy formula is a fundamental tool for comprehending the energy linked to the motion of an object. The formula representing the concept of kinetic energy is expressed as The variable KE in this equation denotes the object's kinetic energy. The variable "m" denotes the object's mass, whereas "v" signifies its velocity (Ferrari & Wunsch, 2009).

$$K.E = \frac{1}{2}mv^2$$

The equation demonstrates that the magnitude of kinetic energy is directly correlated to both the mass and the square of the velocity of an object. This implies that an increase in either mass or velocity will result in a corresponding increase in kinetic energy. Additionally, it is worth noting that the relationship between the kinetic energy and speed is of particular significance because velocity is raised to the power of two in the corresponding formula. This suggests that even minor velocity alterations can substantially influence an entity's kinetic energy (Anderson et al., 2010).

To determine the kinetic energy of an object, it is necessary to possess knowledge of the object's mass and velocity, which can be utilized in applying the provided formula. To calculate the object's kinetic energy, square the velocity and multiply it by half of the object's mass. The resultant quantity will provide the kinetic energy of the object (Sayvetz, 1939).

### 2.2.1 Units of Kinetic Energy

Understanding and quantifying kinetic energy requires units.

J is the SI unit for kinetic energy. 1 kg.m2/s2 = 1 Joule.

This unit represents the kinetic energy of a 1-kilogram object traveling at 1 m/s.

The CGS system uses erg to measure kinetic energy. A one-gram item moving at one centimeter per second has one erg of kinetic energy. The Joule and erg, derived from length, time, and mass, standardize kinetic energy measurement across measurement platforms (Ngamroo & Vachirasricirikul, 2013).

### 2.2.2 Examples of Kinetic Energy

- A river that is flowing at a specific velocity demonstrates kinetic energy due to the combination of its mass and velocity.
- The kinetic energy of an asteroid approaching Earth is substantial due to its mass and velocity.

- iii. In the context of aviation, it is observed that an aircraft possesses a greater amount of kinetic energy due to its significant mass and high velocity.
- The act of ambulation encompasses the coordinated motion of an individual's lower extremities and torso, leading to the acquisition of kinetic energy (Klots, 1976).
- When a spherical object is projected with a certain velocity, it acquires kinetic energy as it traverses through the atmosphere. The magnitude of its kinetic energy is contingent upon the velocity of the throw.
- Objects that descend towards the Earth gather kinetic energy due to the gravitational force acting upon them. The kinetic energy of an impact is directly proportional to the mass and velocity of the object involved.
- The kinetic energy of a truck in motion is higher than that of a car moving at an equivalent velocity. This is because the truck possesses a substantially greater mass than the car (Yamada, 2015).

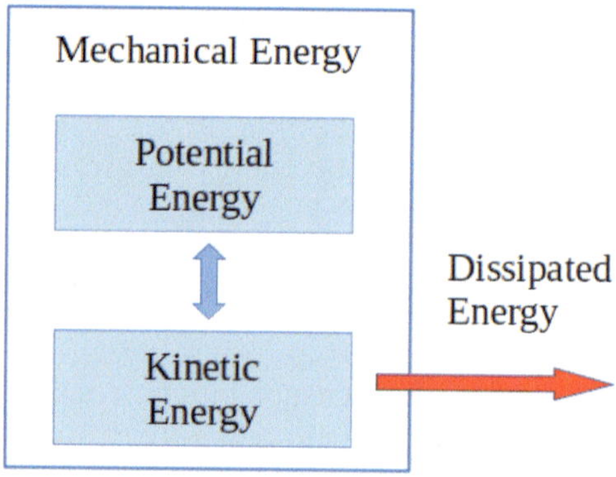


**Figure 2.7.** Illustration of examples of kinetic energy (Source: Ramirez H, Le Gorrec Y. An Overview on Irreversible Port-Hamiltonian Systems. Entropy. 2022; 24(10):1478. https://doi.org/10.3390/e24101478, Creative Commons License).

### 2.2.3 Types of Kinetic Energy

Various forms of kinetic energy can be observed in diverse contexts. Several common types can be identified:

### 2.2.3.1 Translational Kinetic Energy

This pertains to the energy linked to the translational motion of an object. The linear motion of objects characterizes the most commonly encountered form of kinetic energy (Celli et al., 2000).

### 2.2.3.2 Rotational Kinetic Energy

Objects that are in a state of rotation or spinning around an axis demonstrate this particular form of kinetic energy. Illustrative instances encompass spinning tops, wheels, or machinery that undergoes rotation.

### 2.2.3.3 Vibrational Kinetic Energy

The concept of vibrational kinetic energy pertains to the oscillatory movement exhibited by particles within a given system. Vibrating strings, particles, or atoms in a state of vibration have been observed (Császár & Handy, 1995).

### 2.2.3.4 Thermal Kinetic Energy

The concept of thermal kinetic energy pertains to the stochastic movement exhibited by particles within a given material. The phenomenon under consideration is accountable for regulating the thermal conditions and facilitating the exchange of thermal energy within a given substance. The relationship between the velocity of particles and their thermal kinetic energy is such that an increase in particle speed leads to a corresponding increase in thermal kinetic energy (Straede et al., 1987).

### 2.2.3.5 Electrical Kinetic Energy

Electrical kinetic energy pertains to the motion of charged particles, specifically electrons, within the context of an electric current. It is employed in diverse electrical apparatus and serves as a means of energy transmission.

The various forms of kinetic energy exemplify the extensive array of phenomena in which energy is linked to the movement of objects or particles. Distinctive attributes are exhibited by each form of kinetic energy, thereby contributing to various domains within the field of physics and everyday societal applications (Zhou et al., 2022).

## 2.3 POTENTIAL ENERGY

It is widely recognized that an object can store energy due to its spatial arrangement. In the context of a bow and arrow, drawing the bow results in the accumulation of potential energy, which transforms into kinetic energy upon release (Schlegel, 2003).

Likewise, in the context of spring, when it undergoes displacement from its state of equilibrium, it acquires a certain quantity of energy that becomes evident through the stress experienced in hands as they are stretched. Potential energy can be defined as a type of energy that arises from an object's change in position or condition. Upon comprehending the concept of potential energy and its corresponding definition, it is now appropriate to delve into the formula, unit, and illustrative examples of this fundamental concept (Braams & Bowman, 2009).

The calculation of potential energy is contingent upon the magnitude of the force exerted on the two entities involved. The formula for the gravitational force is (Sato, 1955):

$$W = m \times g \times h = mgh$$

Where,

- m is the mass
- g is the gravitational acceleration
- h is the height

### 2.3.1 Types of Potential Energy

Potential energy, alongside other forms, is considered one of the primary manifestations of energy. There exist two primary categories of potential energy:

- Gravitational Potential Energy
- Elastic Potential Energy

#### *2.3.1.1 Gravitational Potential Energy*

The concept of gravitational potential energy refers to the energy that an object possesses when it is elevated to a specific height in opposition to the force of gravity (Lindsey, 2014).

Consider an object with a mass denoted by the symbol "m". Located at an elevation h above the ground, as illustrated in the diagram. It is well-established that the force necessary to lift an object is equivalent to the product of its mass and acceleration due to gravity, commonly referred to as the object's weight (Jones et al., 1996).

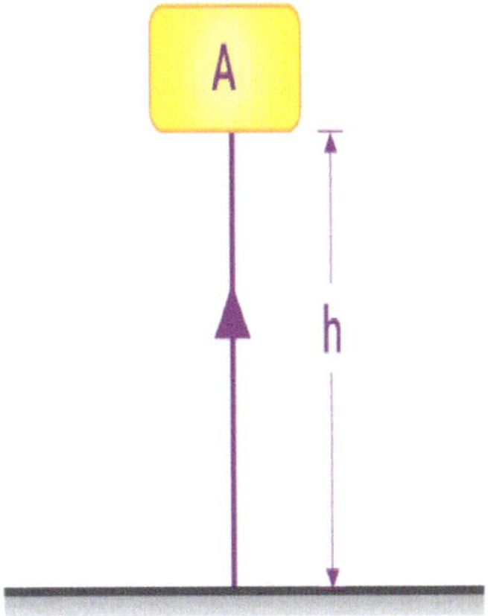


**Figure 2.8.** Illustration of an object with mass (m) placed at height (h) from the ground (Source:, Creative Commons License).

When an object is lifted in opposition to the force of gravity, a certain quantity of work (W) is performed on the object.

The equation for calculating the work done on an object is given by the product of the applied force and the object's displacement (Ghosh et al., 2006).

So,

$$W = m \times g \times h = mgh$$

- According to the principle of energy conservation, the energy gained by an object can be determined by the equation m×g×h, where m represents the object's mass, g represents the acceleration due to gravity, and h represents the height. In this particular scenario, the energy the object gains is equivalent to the potential energy, denoted as E (Hatzfeld et al., 1997).

It is imperative to acknowledge that the gravitational energy is not contingent upon the distance traversed by the object but rather the displacement, specifically the disparity between the object's initial and final heights. Therefore, the trajectory by which the object has attained its elevation is not considered. In the illustration above, it can be observed that the gravitational potential energy of blocks A and B will be equivalent (Flesch et al., 2007).

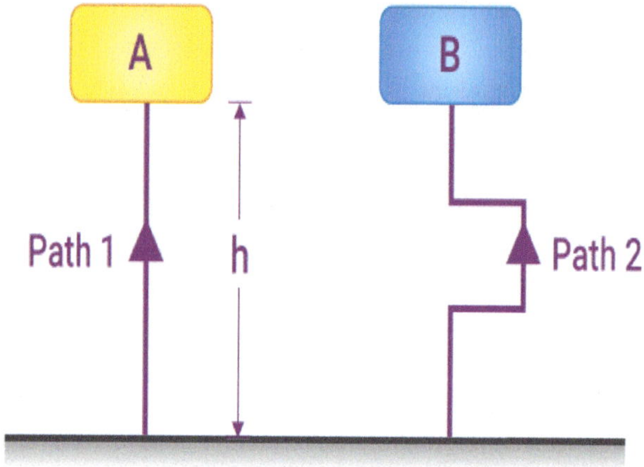


**Figure 2.9.** Illustration of different paths adopted to reach an object with mass (m) placed at height (h) from the ground (Source: Energy Education, Creative Commons License).

### *2.3.1.2 Elastic Potential Energy*

Elastic potential energy accumulates within objects that can undergo compression or extension, exemplified by rubber bands, trampolines, and bungee cords. The greater the extent to which an object can undergo deformation, the higher its potential energy due to elasticity. Numerous objects are purposefully engineered to retain elastic potential energy, exemplified by the subsequent items:

A coiled elastic band that provides propulsion for a miniature aircraft. An archer's taut bow A flexed diver's board before a diver's descent coil spring utilized in a wind-up clock.

Objects that can store elastic potential energy generally exhibit a high elastic limit. Nevertheless, it is important to note that all elastic objects possess a specific threshold for the load they can sustain. Once an object is deformed beyond its elastic limit, it will cease to exhibit the ability to revert to its initial shape.

The calculation of elastic potential energy can be determined by employing the subsequent formula (Liu et al., 2015):

$$U = \frac{1}{2}kx^2$$

Where,

- U is the elastic potential energy
- k is the spring constant
- x is the stretch length

## 2.4 THERMAL ENERGY

Thermal energy is a form of energy that is produced as a result of an increase in temperature. The relationship between thermal energy and an object's temperature change is one of direct proportionality. Heat is a manifestation of thermal energy. The thermal energy of a substance increases proportionally with its temperature. The increase in thermal energy resulting from elevated temperatures can be attributed to the heightened kinetic activity of atoms and molecules within the substance (Sharma et al., 2009).

Occasionally, when the temperature reaches sufficiently elevated levels, the molecular constituents of a substance exhibit sufficient kinetic energy to overcome the intermolecular forces that bind them together, resulting in their dissociation and subsequent departure from the substance. Remarkably, the physical states of matter are contingent upon the amount of thermal energy present. The alteration in thermal energy affects the phase transitions of matter. The augmentation of thermal energy facilitates the transition of a substance from a solid state to a liquid state and, subsequently, from a liquid state to a gaseous state. This phenomenon can be attributed to the rise in the internal energy of the molecules and atoms (Alva et al., 2017).

### 2.4.1 Types of Thermal Energy

All substances are comprised of minuscule particles known as molecules and atoms. Particles are perpetually in motion, exhibiting movement and vibration even at a consistent temperature. Furthermore, the collective sum of their kinetic energies in every direction equals zero. Elevating the temperature augments the molecular kinetic energy, thereby inducing alterations in the state of matter. The thermal energy of matter can be increased through three distinct mechanisms: conduction, convection, and radiation. The following section provides a comprehensive discussion of various methods of heat transfer (Tian & Zhao, 2013):

#### *2.4.1.1 Conduction*

The process of thermal energy transfer is facilitated in the case of conduction, as it involves the transfer of energy between molecules through vibrational motion. Molecules remain relatively stationary within their positions, yet their rapid oscillations facilitate efficient energy transfer. Heat, a manifestation of thermal energy, facilitates the transfer of energy between bodies without any external impediments. Even in a state of thermal equilibrium, molecules, and atoms exhibit kinetic and potential energies in all directions, although their overall impact is null.

When thermal energy is transferred to particles, they undergo vibrational motion within their respective positions and subsequently transfer the excess energy to adjacent particles. In this manner, thermal energy is transferred from one location to another. Thermal energy transfer typically occurs within the solid phase of matter (Shi et al., 2022).

#### *2.4.1.2 Convection*

In the liquid state of matter, heat transfer occurs layered. In the context of liquids, the movements of molecules are not subject to any constraints or limitations.

When subjected to heat, the molecules of the liquid close to the heat source tend to migrate towards regions characterized by lower temperatures. In this process, a fluid undergoes the development of a current, wherein the upward movement of the hot current is accompanied by the influx of the cold current into the vacated space. The procedure above persists until a state is reached wherein the temperature of the liquid is uniformly distributed (Seddegh et al., 2017).

#### *2.4.1.3 Radiation*

In the gaseous phase, the molecular motion is characterized by a high degree of randomness, allowing molecules to exhibit unrestricted movement in any direction. Energy transfer occurs within the waveform in the context of radiation. The waves above are classified as electromagnetic waves, serving as a medium for transmitting energy between molecules.

Thermal heat transfer through radiation does not necessitate the presence of a medium, unlike conduction and convection methods, which rely on a medium to facilitate the transfer of heat or energy. It is worth noting that these can traverse through a vacuum. Hence, the ease of energy transfer is greater in the case of radiation when compared to conduction and convection (Biehs, 2007).

## 2.4.2 Mathematical Representation for Transfer of Thermal Energy

### *2.4.2.1 Conduction*

The process of thermal energy transfer In the context of the conduction method, the heat transfer rate exhibits a direct proportionality to both the temperature difference between the hot and cold surfaces and the area of thermal transfer. The relationship between it and the thickness of the body is inversely proportional. Mathematically,

$$Q = \frac{KA(T_{hot} - T_{cold})}{d}$$

Where Q is the rate of heat transfer

A is the area through which heat transfers

$T_{hot}$ is the temperature of the hot side

$T_{cold}$ is the temperature of the cold side

d is the thickness and

K is the constant of proportionality, known as thermal conductivity (Kumarasamy et al., 2017).

2.4.2.2 Convection

This specific heat transfer mechanism entails consecutive heat transfer across distinct layers. In the given context, the heat transfer rate exhibits a direct relationship with the temperature difference between the hot and cold regions and the surface area encompassed in the heat transfer mechanism. Mathematically,

$$Q = hA(T_s - T_i)$$

Where Q is the rate of heat transfer

A is the area

$T_s$ is the liquid temperature close to the hot plate

$T_i$) is the liquid temperature at the large other end

h is the coefficient of convective heat transfer (Vogel & Thess, 2019).

### *2.4.2.3 Radiation*

The Stefan-Boltzmann law of radiation determines the heat transfer rate in radiation heat or thermal transfer. Based on this legislation, the magnitude of heat transfer is directly correlated with the fourth power of the disparity in temperature between the source and the surrounding environment, as well as the object's surface area. This observation indicates that the thermal transfer via radiation is significantly influenced by the disparity in temperature between the surface and the source. Mathematically,

$$Q = S\dot{o}A(T_s - T_{0)}$$

Where Q is the rate of heat transfer

A is the surface area of the body

ϵ is the emissivity of the body/material

$T_s$ is the temperature at the source side

$T_0$ is the temperature of the object or surroundings side

σ is known as the Stefan-Boltzmann constant, and its value is equal to $5.6703\times 10^{-8}\,\mathrm{W}\,m^{-2}K^{-4}$

The equation above introduces the concept of emissivity. The emissivity of an object refers to its capacity to emit thermal radiation efficiently. A perfect black body is characterized by an emissivity value of 1, indicating its ability to emit and absorb radiation at all wavelengths. Conversely, a perfect mirror has an emissivity value of 0, signifying its capacity to reflect radiation without emission or absorption. The emissivity of actual objects lies within the range defined by these two values (Hsiao, 2017).

## 2.5 CHEMICAL ENERGY

Chemical energy refers to the latent energy stored within atoms' molecular structure. The process of breaking chemical bonds necessitates the input of energy, whereas the formation of new chemical bonds results in the release of energy. The stability of a bond is directly proportional to the amount of energy released during its formation. The stability of these bonds arises from the requirement of an equal amount of energy input to disrupt them, exemplifying the principle of energy conservation. Breaking or forming bonds between atoms is called a chemical reaction (Meng et al., 2016). There exists a multitude of energy-rich molecules on Earth, which possess the capability to engage in chemical reactions that result in the release of usable energy. Oxygen, in the form of O2, is the predominant energy-rich molecule found on Earth. It consists of two oxygen atoms chemically bonded together. These molecules undergo chemical reactions with other molecules, resulting in the release of energy. Most carbon-containing molecules, particularly those that also contain hydrogen and oxygen, have the propensity to undergo a chemical reaction with oxygen, resulting in the formation of carbon dioxide. The combustion process involves burning wood and other materials in the presence of oxygen. The combustion of wood converts its stored chemical potential energy into radiant heat energy. Numerous alternative fuels can undergo comparable combustion reactions (Yu et al., 2015).

Fossil fuels, such as petroleum, coal, and natural gas, represent naturally occurring molecules with a high energy content. The substances in question are macroscopic hydrocarbon compounds derived from ancient flora and fauna, which underwent biological processes, subsequent death, and burial over an extensive span of hundreds of millions of years (for further details on the carboniferous period, please refer to the provided external link). Fossil fuels are extracted through mining, drilling, and pumping processes to serve as a source of energy for various applications, including transportation and electricity production. Fossil fuels are utilized to harness the chemical energy that they contain. Non-renewable resources are characterized by their finite nature and limited quantity (Aslam et al., 2018).

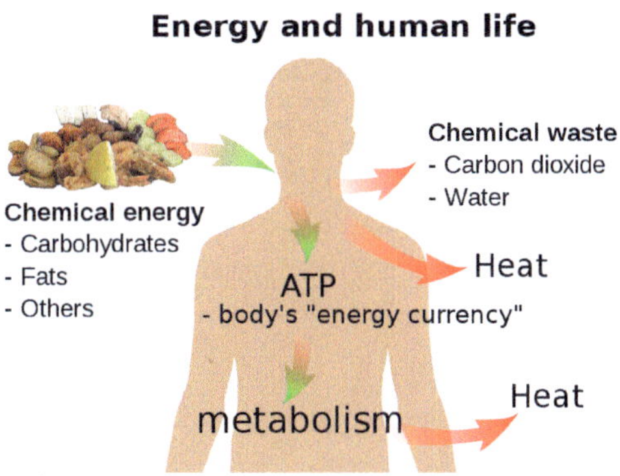


**Figure 2.10.** Representation of different types of chemical energy (Source: By Mikael Häggström - All images are of Public Domain license.File:Pikilia.JPG, Public Domain, https://commons.wikimedia.org/w/index.php?curid=6343781, Creative Commons License).

Biofuels are considered viable substitutes for fossil fuels, as they harness energy derived from organisms that are either alive or have recently perished. In addition to their practical applications, these resources possess utility regarding the chemical energy they contain, rendering them valuable for various purposes. Furthermore, their renewability further enhances their significance as sustainable sources. Hydrogen serves a dual purpose as both a fuel source and an energy currency, as discussed in the literature on hydrogen as an energy currency. The substance in question can undergo combustion similar to fossil fuels, liberating its stored chemical potential energy. However, unlike hydrocarbons, the combustion of this substance does not result in the production of carbon dioxide (CO2) (Ballardini et al., 2001).

### 2.5.1 Examples of Chemical Energy in Everyday Life

Presented below are several instances of chemical energy in our daily lives.

- It is widely recognized that plants rely on solar energy to facilitate the synthesis of sugar from water and carbon dioxide. The cohesion of sugar, carbon dioxide, and water is facilitated by forming chemical bonds that maintain the integrity and stability of these substances.
- Consider, for instance, that all sugars consist of carbon, hydrogen, and oxygen atoms interconnected by multiple chemical bonds. Nevertheless, the bonding of these atoms does not occur spontaneously, necessitating the input of energy to facilitate their cohesion (Tolmachev et al., 2015).
- Plants utilize solar energy to synthesize sugar by combining carbon, hydrogen, and oxygen atoms into a cohesive molecular structure. This exemplifies a notable instance of energy conversion, whereby energy transforms one form to another. In this particular instance, solar energy is converted into chemical energy.
- In energy storage, dry wood serves as a repository for chemical energy. During combustion, the chemical energy is released and subsequently transformed into light and thermal energy. It should be noted that wood undergoes a transformation into ashes, forming a distinct substance.
- The consumption of food serves as a suitable illustration of the release of stored chemical energy during the process of

digestion. The molecular constituents of food undergo a process of fragmentation, resulting in the formation of smaller entities. Chemical reactions occur when the bonds between atoms are disrupted, leading to the formation of new compounds. Simultaneously, the process of oxidation occurs rapidly as the bonds are broken (Cooper & Klymkowsky, 2013).

### 2.5.2 Characteristics of Chemical Energy

Chemical energy is liberated upon a chemical reaction, typically as thermal energy. The reactions above are commonly referred to as exothermic reactions. Reactions that require an input of heat to proceed have the potential to store a portion of this energy as chemical energy within the newly created bonds. The body undergoes a process of converting food's chemical energy into thermal energy and mechanical work. At a power station, chemical energy contained in coal is converted into electrical energy. The battery harnesses its chemical energy to facilitate the process of electrolysis, thereby enabling electrical power generation (Tang et al., 2015).

## 2.6 NUCLEAR ENERGY

Nuclear energy is contained within an atom's nucleus or central region. Atoms are minuscule entities that constitute the entirety of matter in the cosmos, and the cohesive force responsible for maintaining the integrity of the nucleus is known as energy. The nucleus of an atom contains a significant quantity of energy. The force responsible for binding the nucleus is called the "strong force." Using nuclear energy for electricity generation necessitates the initial liberation of said energy from the atomic structure. During nuclear fission, the atomic nuclei undergo division, resulting in the liberation of energy (Zinkle & Was, 2013).

A nuclear reactor, also called a power plant, is a complex assemblage of machinery designed to regulate the process of nuclear fission to generate electrical energy. Nuclear reactors employ uranium pellets as the fuel source for facilitating nuclear fission reactions. Within the context of a nuclear reactor, the process involves inducing the fission of uranium atoms. During nuclear fission, the atomic nuclei divide, emitting minuscule particles known as fission products. Fission products induce the subsequent splitting of additional uranium atoms, thereby initiating a self-sustaining chain reaction. The thermal energy generated from this series of reactions is responsible for heat production (Lenzen, 2008).

The thermal energy generated through nuclear fission elevates the temperature of the cooling medium within the reactor. Typically, water is a cooling agent; however, certain nuclear reactors employ alternative substances, such as liquid metal or molten salt, for this purpose. The coolant, which undergoes thermal energy release through nuclear fission, generates steam. The process involves converting steam energy into mechanical energy through the rotation of turbines, which are devices consisting of wheels propelled by the force exerted by a flowing current. The turbines are responsible for propelling generators, which generate electrical power (Koning & Rochman, 2008).

Rods composed of nuclear poison can regulate the magnitude of electrical output. Nuclear poisons refer to specific substances, such as xenon, which can absorb some fission products generated during nuclear fission. A greater number of rods containing nuclear poison in the context of a chain reaction will result in deceleration and increased reaction regulation. Removing the rods will facilitate a more potent chain reaction and increase electrical power generation (Hubbert, 1956).

As of 2011, approximately 15 percent of global electricity production is derived from nuclear power plants. The United States possesses over 100 reactors, despite the primary sources of its electricity generation being derived from fossil fuels and hydroelectric power. Countries such as Lithuania, France, and Slovakia predominantly rely on nuclear power plants to generate their electricity (Petrescu et al., 2016).

**Did You Know?**

In 1938, German physicists Otto Hahn and Fritz Strassmann, along with Austrian physicist Lise Meitner, discovered nuclear fission.

### 2.6.1 Nuclear Food: Uranium

Uranium is the primary fuel source utilized for generating nuclear energy on a global scale. This phenomenon can be attributed to the relatively facile fission of uranium atoms. Uranium is a prevalent chemical element widely distributed within the Earth's crust, occurring in various rock formations across the globe. Nevertheless, the particular variety of uranium utilized for generating nuclear power, known as U-235, is characterized by its scarcity. The isotope U-235 constitutes a fraction of less than one percent of the total uranium present on Earth (Garten, 1978). While some of the uranium utilized by the United States is sourced domestically, most are acquired through imports. The United States acquires uranium from several countries, including Australia, Canada, Kazakhstan, Russia, and Uzbekistan. After uranium mining, it becomes necessary to separate it from other minerals. Before utilization, the substance must undergo processing (Thomas & Gates, 1999).

The importation of nuclear fuel, such as uranium or plutonium, is restricted to nations that are signatories of the Nuclear Non-Proliferation Treaty (NPT) due to the dual-use nature of these materials, which can be utilized for both nuclear weapons and nuclear reactors. The treaty promotes the peaceful utilization of nuclear fuel while concurrently imposing restrictions on the proliferation of nuclear weapons.

A conventional nuclear reactor typically consumes approximately 200 metric tons of uranium annually. Sophisticated procedures facilitate the re-enrichment or recycling of certain quantities of uranium and plutonium. This practice diminishes the quantity of mining, extraction, and processing activities that are required (Hamada & Ogino, 2012).

## 2.6.2 Nuclear Energy and People

Nuclear energy generates electrical power that can be effectively utilized for residential, educational, commercial, and medical purposes. The initial nuclear reactor was designed to generate electricity near Arco, Idaho. The Experimental Breeder Reactor achieved self-sustaining power generation in the year 1951. The inaugural nuclear power plant aimed to supply energy to a local community was established in Obninsk, Russia 1954 (Jun et al., 2010).

The construction of nuclear reactors necessitates advanced technological capabilities and access to uranium or plutonium, which are exclusively available to countries that have ratified the Nuclear Non-Proliferation Treaty. Due to various factors, the developed world is the primary location for most nuclear power plants. Nuclear power plants generate sustainable and environmentally friendly energy. These entities lack air pollution and do not emit greenhouse gases. Wind turbines can be constructed in both urban and rural settings without causing significant environmental changes in their surroundings (De Groot & Steg, 2010).

The steam used to drive the turbines and generators undergo a recycling process. The cooling process occurs within a distinct infrastructure known as a cooling tower. The steam undergoes a phase transition, returning to its liquid state, thereby enabling its reuse to generate additional electrical power. The surplus steam is effectively reintroduced into the atmosphere, where its impact is minimal as it transforms into harmless water vapor (Roeser, 2011).

Nevertheless, one of the consequences of nuclear energy is producing radioactive material. Radioactive material refers to a compilation of atomic nuclei that exhibit instability.

The dissipation of energy from these nuclei has the potential to impact various materials in their vicinity, including living organisms and the surrounding environment. Radioactive substances possess a high degree of toxicity, leading to severe burns and elevated susceptibility to the development of malignancies, hematological disorders, and osteoporosis (Van der Pligt et al., 1986).

Radioactive waste refers to the residual materials that remain following the functioning of a nuclear reactor. Radioactive waste primarily comprises protective clothing utilized by workers, tools, and other materials that have come into contact with radioactive dust. Its enduring nature characterizes it. Certain materials, such as clothing and tools, have the potential to retain their radioactive properties for an extended duration spanning thousands of years. Disposable materials are subject to government regulations to prevent contamination of other entities (Routley & Routley, 1978).

The fuel and rods utilized in nuclear reactors possess a high level of radioactivity. The utilized uranium pellets necessitate containment within specialized receptacles resembling expansive swimming pools. The primary function of water in this context is to facilitate fuel cooling and provide insulation to prevent direct contact between the external environment and any potential radioactivity. Certain nuclear power plants employ a method of storing their spent fuel in above-ground dry storage tanks.

The contentious nature of radioactive waste storage facilities has engendered significant controversy within the United States. Over an extended period, the government has been engaged in the strategic planning of a substantial nuclear waste facility close to Yucca Mountain, Nevada, as an illustrative example. The plan was met with opposition from environmental organizations and local community members. Concerns were raised regarding the potential leakage of radioactive waste into the water supply and the surrounding environment of Yucca Mountain, approximately 130 kilometers (80 miles) from the densely populated urban region of Las Vegas, Nevada. The government initiated an investigation into the Yucca Mountain site in 1978 but ceased its efforts to develop a nuclear waste facility at this location in 2009 (Sonnberger et al., 2021).

### 2.6.3 Applications of Nuclear Energy

- Nuclear technology
- Nuclear medicine
- Industrial uses
- Agricultural uses
- Environmental uses
- Biological Experimentations
- Medical diagnosis and treatments
- Scientific Investigations and Engineering Projects
- Neutron Activation Analysis

### 2.6.4 Future of Nuclear Energy

Nuclear reactors employ the process of fission, which involves the division of atomic nuclei to generate energy. Nuclear energy can also be generated via fusion, which involves combining atomic nuclei or joining them together. For example, the sun perpetually engages in the process of nuclear fusion, wherein hydrogen atoms combine to produce helium. The viability of life on Earth is contingent upon the sun, thereby establishing nuclear fusion as a fundamental contributor to the sustenance of life on our planet (Hill, 2008).

**Did You Know?**

In the late 19th century, scientists such as Henri Becquerel, Marie Curie, and Pierre Curie made significant discoveries related to radioactivity.

Nuclear power plants cannot effectively and dependably generate energy through nuclear fusion. The feasibility of utilizing this process to generate electricity remains uncertain. Nuclear engineers are currently engaged in extensive research on nuclear fusion, primarily due to this process's anticipated safety and cost-effectiveness (Brook & Bradshaw, 2015).

## 2.7 ELECTRICAL ENERGY

Electrical energy refers to the capacity of an electrical circuit to generate work through the initiation of a specific action. This phenomenon can manifest in various modalities, including thermal, electromagnetic, mechanical, electrical, etc. Electrical energy can be generated through various means, such as batteries, power plants, dynamos, and photovoltaics. It can also be stored for future use by employing fuel cells, batteries, capacitors, or magnetic fields, among other methods. Therefore, electrical energy can be generated or stored (Gutowski et al., 2006).

In accordance with our previous instruction in science courses, it is widely acknowledged that the fundamental principle known as "The Law of the Conservation of Energy" posits that energy cannot be generated nor annihilated but rather undergoes conversion.

However, for energy to be effectively utilized, converting it from one form to another is necessary. As an illustration, a motor can transform electrical energy into mechanical or kinetic energy. In contrast, a generator can convert kinetic energy back into electrical energy to supply power to a circuit (Dunn et al., 2011).

Electrical machines possess the capability to transform energy from one form to another through the performance of mechanical work. An additional illustration can be found in the form of a lamp, light bulb, or LED (light emitting diode), all of which can transform electrical energy into light and thermal energy. Electrical energy exhibits a high degree of versatility due to its inherent capacity for seamless conversion into various alternative forms of energy (Kondoh et al., 2000).

It is necessary to exert a certain amount of work to facilitate the movement of electrons and generate a continuous flow of electric current within a circuit. This work entails the displacement of electrons over a given distance within a wire or conductor. The energy generated by the work performed is stored within the flow of electrons. Therefore, the term "Work" is employed to denote the phenomenon of energy transformation. Consequently, it can be asserted that Work and Energy are synonymous, given that energy can be defined as the capacity to perform work. It is important to acknowledge that the concepts of work done and energy transferred apply to mechanical and thermal systems, just like electrical systems. The interchangeability of mechanical, thermal, and electrical energies is the underlying reason for this phenomenon (Whittingham, 2008).

A cell comprises two terminals: a negative terminal and a positive terminal. An abundance of electrons characterizes the negative terminal, while the positive terminal exhibits a deficit of electrons. Designate the positive terminal as A and denote the electrical potential at A as V(A). Likewise, the terminal denoted as B is designated as the negative terminal, and the corresponding electrical potential at B is represented by the variable V(B). The direction of electric current is from point A to point B, resulting in a higher voltage at point A compared to point B.

The calculation of the potential difference between points A and B is determined by (Hameer & van Niekerk, 2015):

$V=V(A)-V(B)>0$

Since the potential energy of charge, Q at A is Q V(A), and at B, it is Q V(B),

$\Delta U_{pot}$ = Final– Initial potential energy

= D Q [V (B)-V (A)]=-D Q V

=- IVDt (Since I=DQ/Dt)

Considering the inclusion of the system's kinetic energy, it would undergo alteration if the charges within the conductor were to move without experiencing any collisions. The purpose of this is to maintain the system's overall energy at a constant level. Therefore, based on the principle of conservation of total energy:

DK=-DUpot

Or

DK=IVDt>0

Therefore, within the context of the electric field, when charges exhibit unhindered mobility across a conductor, their kinetic energy experiences a corresponding augmentation during their displacement.

The resulting energy is distributed among the atoms involved when the charges undergo collision. As a result, the atomic vibrations intensify, leading to the conductor's thermal energy transfer and subsequent heating. Consequently, a certain quantity of energy is dissipated as thermal energy within a practical conductor (Ferreira et al., 2013).

## 2.7.1 Units of Electrical Energy

The fundamental unit of electrical energy is the joule, equivalent to one watt-second. The unit of electrical energy, known as the joule, is the amount of energy transferred when a current of one ampere passes through a circuit for one second while a potential difference of one volt is applied. The kilowatt-hour (kWh) is the recognized commercial unit for measuring electrical energy, alternatively referred to as the Board of Trade unit (B.O.T) (Batih & Sorapipatana, 2016).

### 2.7.2 Examples of Electrical Energy

There exist several instances of electrical energy, namely:

- In the context of a car battery, a chemical reaction occurs, leading to the generation of an electron that possesses sufficient energy to facilitate its movement within an electric current. The mobile charges within the car's circuits facilitate the transmission of electrical energy.
- Lightning, which occurs in the context of a thunderstorm, serves as an illustration of electrical energy. The phenomenon commonly referred to as lightning can be understood as the electricity discharge within the Earth's atmosphere.
- Electric eels can produce electrical energy, which they employ to defend against potential predators (Demiroren & Yilmaz, 2010).

## 2.8 ELECTROMAGNETIC ENERGY

Electromagnetic energy is the form of radiant energy that propagates through space through waves, moving at the constant speed of light. It can be characterized as radiant energy, electromagnetic energy, luminosity, or the propagation of radiation. Electromagnetic radiation can facilitate the transfer of thermal energy. Electromagnetic waves serve as carriers of thermal energy, radiant energy, or optical radiation, propagating through either a vacuum or a medium, facilitating the transfer of these forms of energy between distinct locations. The phenomenon under consideration is classified as electromagnetic energy (Maier & Atwater, 2005).

The discovery of electromagnetic radiation is attributed to James Clerk Maxwell, a prominent physicist of the 19th century. His research and subsequent findings significantly impacted the development of the field now recognized as quantum mechanics. Regarding its operational mechanism, electromagnetic energy or radiation can be conceptualized as analogous to a conventional ocean wave. In this symbolic context, the radiation is likened to the element of water. Electromagnetic waves can be likened to ocean waves, exhibiting wave-like properties. In the case of electromagnetic waves, energy propagation occurs through the oscillation of electric and magnetic fields. This analogy suggests that electromagnetic energy is generated through the transfer of energy from the source to the surrounding medium, similar to how water is transported from the central regions of the ocean toward the shoreline by ocean waves (Brevik, 1979).

The energy in question is most effectively demonstrated by the magnitude of power required to transport substantial volumes of water over considerable distances. Electromagnetic energy transfer and generation exhibit a higher level of complexity (Quinten et al., 1998).

### 2.8.1 How Do Electromagnetic Waves Work

Electromagnetic energy comprises dynamic magnetic and electric fields that facilitate electromagnetic energy transfer. Positive charges generate electric fields, resulting in the formation of a charged region that extends outward. The electric field can be altered by manipulating a charged particle, such as vertically displacing it (Joulain et al., 2005).

The presence of magnetic currents generates magnetic fields. Changes in the magnetic field can manifest when there is an oscillation in the magnetic current. The interaction between magnetic and electric fields results in mutual influence, whereby changes and displacements in one field correspondingly affect the other. The magnetic fields propagate along a horizontal plane, while the electric fields propagate vertically, thereby enabling the polarization of electromagnetic fields (Diddams et al., 2020).

The fundamental constituents of electromagnetic waves are the propagation of electric and magnetic fields, which govern the movement of these waves. A changing magnetic field induces a changing electric field, which in turn induces a changing magnetic field. The outcome manifests as a sequential series of events wherein these domains exhibit oscillatory behavior in mutually perpendicular orientations, generating transverse electromagnetic waves. The propagation of waves occurs through the transmission of radiation particles known as photons, which possess negligible mass and can traverse at the velocity of light. Transverse waves, propelled by magnetic fields and energetic photons, are responsible for the propagation of electromagnetic energy waves. The electromagnetic spectrum refers to the range of frequencies and wavelengths that electromagnetic waves can potentially exhibit (White, 2001).

### 2.8.2 Electromagnetic Spectrum

The electromagnetic spectrum encompasses a continuum of frequencies and wavelengths associated with electromagnetic radiation. Various combinations of wave types and frequencies give rise to distinct manifestations of energy. The electromagnetic frequency can be defined as the rate at which wave crests propagate to a designated location within a given time interval. Frequency can be conceptualized as the occurrence of each wave's crest during its propagation and displacement. The unit of frequency, denoted as Hz, represents one wave cycle per second (DeVany et al., 1968).

The unit of frequency known as the Hertz (Hz) is attributed to its namesake, Heinrich Hertz, a renowned German physicist. Hertz conducted pioneering experiments involving radio waves, which aimed to demonstrate the equivalence between the velocity of these waves and the speed of light or radiation. The discovery had significant implications for the field of electromagnetic energy. The velocity of a wave can be determined by multiplying the wavelength by the frequency. As the frequency of an electromagnetic wave increases, there is a corresponding decrease in its wavelength, increasing the wave's power (Psaltis, 2008).

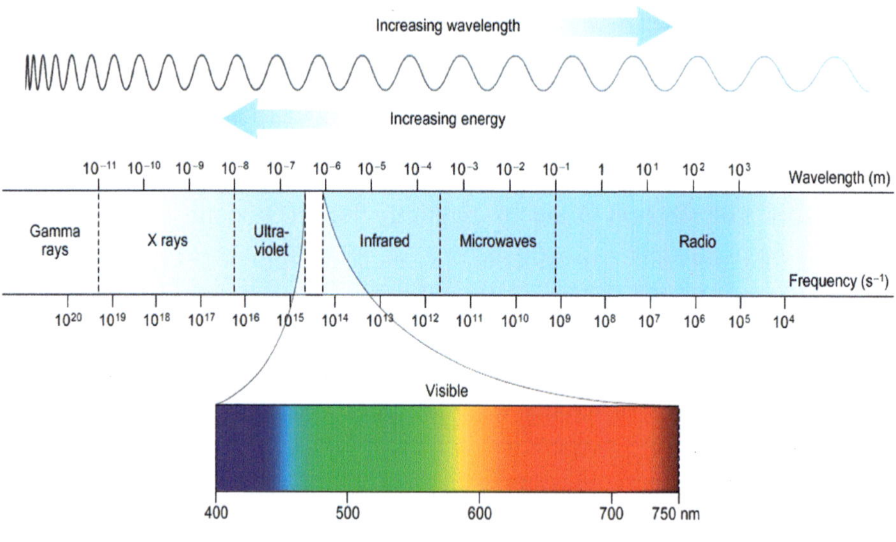


**Figure 2.11.** Illustration of the electromagnetic spectrum (Source: Gartshore A, Kidd M, Joshi LT. Applications of Microwave Energy in Medicine. Biosensors. 2021; 11(4):96. https://doi.org/10.3390/bios11040096, Creative Commons License).

The measurement of electromagnetic wave energy is typically expressed in units of electron volts (eV). This unit signifies the kinetic energy to transfer electrons through a voltage potential. Stated differently, the quantification of energy pertains to the amount of energy required to generate additional waves or crests. The quantity of electromagnetic waves increases as the size of the waves decreases, resulting in a corresponding increase in energy. A greater wavelength corresponds to a decrease in energy and, consequently, a reduction in frequency. Conceptualize the electromagnetic spectrum as a linear, horizontally oriented continuum that is being traversed in a left-to-right manner.

On the left side of the spectrum, there is a decrease in frequency, measured in hertz, and an increase in wavelength. Towards the rightmost extremity of the spectrum, one encounters waveforms of reduced magnitude and increased frequency, measured in hertz. As one traverses the spectrum from one end to the other, the electromagnetic energy assumes greater prominence, corresponding to a heightened intensity in frequency (Kemp & Kemp, 1991).

## 2.8.3 Electromagnetic Radiation across the Spectrum

Spanning the spectrum are seven types of electromagnetic radiation:

### *2.8.3.1 Radio Waves*

Low-frequency radio waves are located at the beginning of the electromagnetic spectrum. Low-frequency radio waves exhibit the longest wavelengths and possess the lowest energy within the electromagnetic spectrum. Their dimensions range from the scale of a football field to surpassing the size of the Earth. Radio waves are utilized for radio broadcasting and telescope technology, enabling the reception of radio signals and observing celestial objects in space (Van de Hulst, 1945).

### *2.8.3.2 Microwaves*

Microwaves exhibit similarities to radio waves in frequency and size; however, they diverge due to the requisite technology for their utilization and the technological capabilities they offer. Microwaves of different types are distinguished by their respective wavelengths. As an illustration, it is worth noting that C-band or medium-sized microwaves can penetrate various atmospheric conditions such as clouds, snow, rain, dust, smoke, or haze, facilitating satellite communication. On the other hand, L-band microwaves find application in the operation of global positioning systems (GPS). Microwaves are instrumental in facilitating the operation of television and cellular communication signals and serve a crucial role in microwave oven functionality (Rosen et al., 2002).

#### 2.8.3.3 Infrared Waves

Infrared waves, alternatively referred to as infrared light or radiation, can be perceptible to human beings through thermal sensation. The electromagnetic spectrum encompasses various regions, including the infrared section. Within this section, there are three distinct subsectors, namely near-infrared, mid-infrared, and far-infrared. The term "far infrared" is often called "thermal infrared" due to its optimal suitability for observing thermal or heat energy. The utilization of infrared electromagnetic energy encompasses various applications, including detecting and observing celestial objects, monitoring and analyzing Earth's temperature patterns, visualizing objects or heat energy through thermal imaging, and facilitating channel switching on televisions via remote control devices (Kay et al., 2009).

#### 2.8.3.4 Visible Light

The visible light spectrum is conventionally situated in the central region of the electromagnetic spectrum. This segment of the electromagnetic spectrum corresponds to the range of wavelengths that are perceptible to the human visual system. All forms of electromagnetic radiation are classified as light; however, the term "visible light" or "visible spectrum" is specifically used to refer to the subset of electromagnetic radiation that the human visual system detects. The visible light spectrum encompasses the range of wavelengths that are perceptible to the human eye, resulting in the formation of a rainbow. Each distinct color within the rainbow corresponds to a specific wavelength, with variations in size. As an illustration, it is worth noting that red is characterized by having the longest wavelengths, while violet is distinguished by possessing the shortest wavelengths (Banerjee et al., 2014).

#### 2.8.3.5 Ultraviolet (UV) Waves

Ultraviolet waves, alternatively referred to as UV light and ultraviolet radiation, occupy the higher frequency range of the electromagnetic spectrum due to their comparatively shorter wavelengths and higher energy levels. Ultraviolet radiation is categorized into various levels of intensity, encompassing near, middle, far, and extreme ultraviolet light. Nevertheless, excessive exposure to ultraviolet light can pose a risk to human health due to its elevated frequency and energy levels.

Ultraviolet (UV) radiation can inflict harm or detriment upon the integumentary system through the induction of sunburn, cellular disruption, and potential alteration of genetic material. The primary reason individuals utilize sunscreen is to safeguard their skin against the sun's ultraviolet (UV) radiation (Perera et al., 2022).

#### *2.8.3.6 X-Rays*

X-rays occupy the penultimate position on the electromagnetic spectrum. These rays exhibit a significantly elevated energy frequency and considerably reduced wavelength size, potentially reaching dimensions as minute as an atom's. The wavelength of X-rays is contingent upon the temperature of the object, whereby higher temperatures correspond to shorter wavelengths, and lower temperatures correspond to longer wavelengths. X-rays are widely recognized for their application in medical imaging, wherein X-ray waves are transmitted through the human body, forming shadows of objects on X-ray films.

Excessive exposure to X-ray waves can pose a risk to the human body. The use of protective equipment by patients undergoing medical X-ray procedures and the practice of X-ray technicians vacating the room during image acquisition is rooted in the following rationale (Kronig, 1926).

#### *2.8.3.7 Gamma Rays*

Gamma rays at the end of the electromagnetic spectrum are characterized by their short wavelengths and the highest energy frequency. Consequently, gamma waves exhibit the highest level of electromagnetic intensity. Gamma rays are produced through various phenomena, such as supernova explosions, black holes, nuclear decay, nuclear reactions, and lightning discharges. According to NASA, the energy generated by these bursts of rays is so immense that it surpasses the sun's total energy output over its entire lifespan within a mere 10-second interval (Magdziarz & Zdziarski, 1995).

### 2.8.4 Is Electromagnetic Energy Safe

The potential hazards linked to electromagnetic waves raise the inquiry regarding the safety of electromagnetic energy. The characteristics of electromagnetic radiation vary across the entirety of the electromagnetic spectrum, encompassing a range of different types of radiation. Ionizing radiation is generated by electromagnetic energy with the highest frequencies, encompassing ultraviolet, X-ray, and gamma-ray waves. Gamma rays present a significant hazard in terms of ionizing radiation, which is generated due to nuclear reactions and incidents. Furthermore, it should be noted that nuclear decay can pose potential health risks associated with ionizing radiation, which can be emitted in the form of gamma rays or X-rays. The potential consequences of exposure to ionizing radiation include the induction of carcinogenic DNA damage, the manifestation of radiation sickness, and the occurrence of mortality. Non-ionizing radiation, in contrast, lacks sufficient energy to present significant radiation concerns or hazards. This refers to radiation emitted by lower-frequency waves, including visible light, microwaves, and radio waves (Osepchuk & Petersen, 2003).

Non-ionizing radiation refers to the type of radiation that individuals are commonly exposed to when using devices emitting electromagnetic waves, such as mobile phones, televisions, computers, power lines, or microwaves. Nevertheless, lower-frequency radiation will gradually transition to increasingly worrisome levels as global warming persists. Solar radiation is directed towards the Earth's surface and reflected into space. Greenhouse gases, which are a form of pollution resulting from emissions, possess the ability to confine radiation within the Earth's atmosphere, thereby engendering the greenhouse gas effect and perpetuating the phenomenon of global warming (Albanese et al., 1994).

### 2.8.5 Importance of Electromagnetic Energy

With the increasing concern over the planet's environmental condition, there is a corresponding imperative to enhance our comprehension of electromagnetic radiation. As the demand for renewable and sustainable energy sources increases, scientists must continue investigating radiation and electromagnetic energy. Furthermore, the ongoing advancement of technology in the form of computers, smartphones, energy-efficient household appliances, and renewable energy sources will continue to be of utmost importance in addressing the escalating demand for connectivity and information in a world experiencing rapid population growth.

The acquisition of knowledge and practical application of electromagnetic energy will enable to sustain our utilization of the electromagnetic waves that serve as our society's primary power source (Thayer et al., 1995).

## 2.9 GRAVITATIONAL ENERGY

Gravitational energy refers to the energy that an object possesses or obtains due to a positional change within a gravitational field. In more precise terminology, gravitational energy can be defined as the form of energy intrinsically associated with the gravitational force or the phenomenon of gravity. When an object with mass (m) is transported from an infinite distance to a location within the gravitational field of a source mass (M) without undergoing acceleration, the work performed in moving the object into the gravitational field is converted into potential energy. This phenomenon is commonly referred to as gravitational potential energy. The symbol Ug is utilized to denote it (Hayward, 1996).

An illustrative instance that aids in comprehending the notion of gravitational potential energy involves the utilization of two pencils, whereby one is positioned on a table surface. At the same time, the other is suspended above the table. It can be asserted that the pencil positioned at a higher elevation possesses a greater amount of gravitational potential energy than the pencil on the table (Deser & Tekin, 2002).

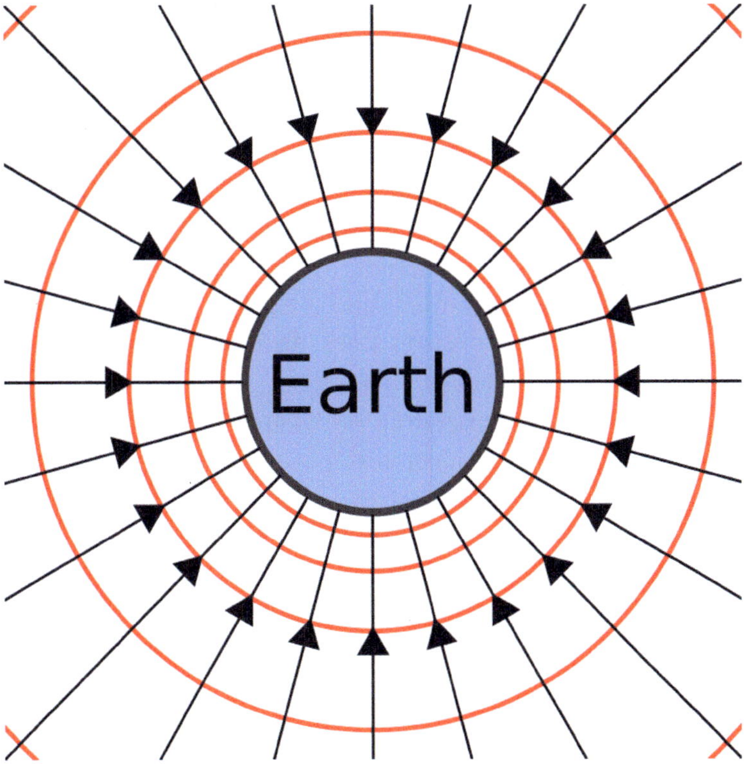


**Figure 2.12.** Illustration of gravitational potential energy (Source: By Sjlegg - Own work, Public Domain, https://commons.wikimedia.org/w/index.php?curid=6737540, Creative Commons License).

The concept of potential energy entails the energy that is stored within a body when it occupies a specific position. When external forces are applied to a body, resulting in a change in its position, the corresponding change in potential energy is equivalent to these forces' work on the body (Horowitz & Perry, 1982).

The work performed under the influence of gravitational force remains constant regardless of the trajectory followed during a displacement, thus establishing the conservative nature of the force. Moreover, it is important to note that all these forces possess inherent potential. The gravitational effect on a celestial object at an infinite distance is negligible, resulting in zero gravitational potential energy. This state is commonly referred to as a reference point.

Electric energy generation in a hydroelectric power plant is sourced from the conversion of gravitational energy. The initial stage of the process involves the exertion of gravitational forces, which result in the downward movement of water towards the lower region of the dam. Consequently, the gravitational potential energy the water possesses is transformed into kinetic energy, which is attributable to the motion of matter. When the fluid flow encounters the blades of a turbine, a portion of its kinetic energy is transformed into rotational energy, resulting in the rotation of the turbine. The turbine is connected to an electric generator, whereby the turbine's mechanical energy is transformed into electrical energy. The electric generator is linked to transmission lines, facilitating the distribution of electric energy to end-users (Johri et al., 1995).

The primary method of generating electricity for consumer use involves the utilization of large-scale generators powered by steam turbines. The efficiency and cost-effectiveness of the system are optimized when the generators consistently generate electricity at a steady rate. Nevertheless, a significant issue arises due to the substantial fluctuation in demand throughout different periods of the day. During periods of low consumer demand, it would be advantageous for a utility company to operate its most efficient generators and store electrical energy for future use.

Regrettably, storing substantial quantities of electric energy generated by a large generator lacks a feasible solution. It is conceivable, nevertheless, to transform the electrical energy into an alternative form and subsequently store it. Subsequently, when electricity is demanded, the secondary energy source undergoes a reconversion process to restore it to its original electrical form. One potential approach involves the utilization of efficient generators during periods of low demand, particularly at night. These generators would power electrically driven pumps, storing water as gravitational energy within a raised reservoir (Maluf, 2005).

The equation for gravitational potential energy is:

$GPE = m \cdot\times g \cdot\times h$

Where,

- m is the mass
- g is the gravitational acceleration
- h is the height

## 2.10 ELASTIC ENERGY

Elastic energy is the potential energy stored by stretching or compressing an elastic object by an external force such as the stretching of a spring. It is equal to the work done to stretch the spring which depends on the spring constant k and the distance stretched (Wilson & Flanagan, 2008).

Many objects are designed specifically to store elastic potential energy, for example:

- The coil spring of a wind-up clock
- An archer's stretched bow
- A bent diving board, just before a diver's jump
- The twisted rubber band which powers a toy airplane
- A bouncy ball, compressed at the moment it bounces off a brick wall.

Objects that are specifically engineered to store elastic potential energy generally possess a considerable elastic limit. Nevertheless, it is important to note that all elastic objects possess a finite threshold beyond which they cannot sustain a load. Once an object is subjected to deformation beyond its elastic limit, it will cease to exhibit the ability to revert to its initial shape. During previous eras, there was a notable prevalence of wind-up mechanical watches that were widely embraced as fashionable accessories, functioning through the utilization of coil springs for power. In contemporary times, wind-up smartphones are not commonly employed due to the absence of materials possessing an elastic limit that is sufficiently high to store elastic potential energy at a density that meets the required threshold (Nia et al., 2016).

Hooke's law states that the force exerted to elongate a spring is directly proportional to the magnitude of the elongation.

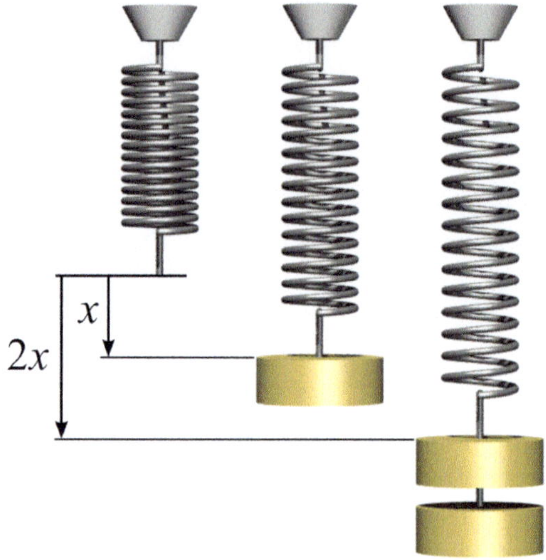


**Figure 2.13.** Illustration of Hooke's Law (Source: By Svjo - Own work, CC BY-SA 3.0, https://commons.wikimedia.org/w/index.php?curid=25398333, Creative Commons License).

In other words, the force required to stretch the spring directly varies with its displacement:

$F=kx$

Wherein,

k = spring constant

x = displacement

The Elastic Potential Energy of the spring is given by (Biewener & Roberts, 2000):

$$P\cdot E = \frac{1}{2}kx^2$$

## 2.11 SOUND ENERGY

Sound energy is a kinetic energy variant arising from the mechanical oscillation of air particles or molecules. The particles collide with adjacent particles, resulting in their subsequent vibrational motion. The propagation of these vibrations occurs in a linear trajectory. Upon reaching our auditory organs, these vibrations are interpreted by our sensory system as auditory stimuli, commonly referred to as sound.

When an auditory stimulus is perceived, it involves the reception of sound waves that propagate through the ear canal and subsequently cause the eardrum to vibrate, analogous to the vibrations observed in a drum head upon being struck. Various auditory stimuli produce distinct vibrations that influence the movement of the tympanic membrane.

The transmission of vibrations occurs from the tympanic membrane through the ossicular chain to the cochlea, a fluid-filled anatomical structure. This process induces surface waves within the cochlea that subsequently interact with hair cells. The perception of high or low-pitched sounds by the brain through the auditory nerve is contingent upon the spatial arrangement of the hair cells within the cochlea. Subsequently, it converts the primary oscillations of the air particles within the auditory wave into audible manifestations that are understandable to us (Marshall, 1994).

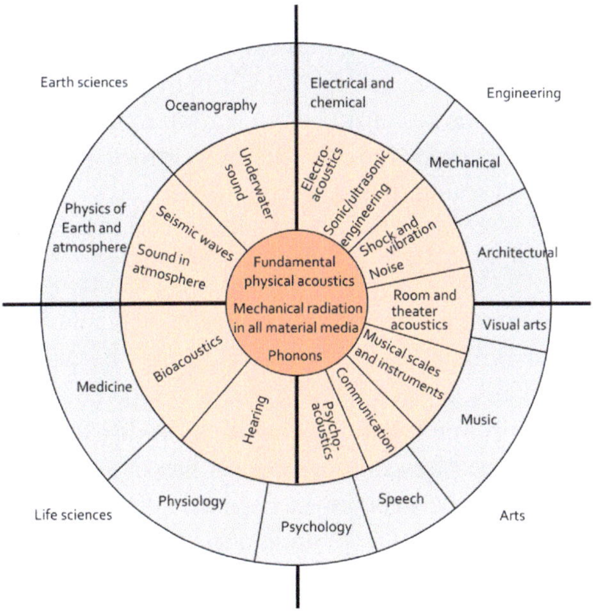


**Figure 2.14.** Representation of examples of sound energy (Source: By Amitchell125 - Own work, based on https://www.acousticbulletin.com/simplifying-acoustics, CC BY-SA 4.0, https://commons.wikimedia.org/w/index.php?curid=102800100, Creative Commons License).

### 2.11.1 How Does Sound Energy Work

The particles vibrate about their position until they reach an equilibrium. The vibrations, and hence sound, are transmitted as a wave, known as a sound wave. The particles vibrate in the same direction as the wave propagates. Such a type of wave is known as a longitudinal wave. The sound wave carries energy and keeps traveling until it loses energy.

Sound requires a medium to propagate. It can travel through air, water, wood, glass, or metal. Unlike light, sound cannot propagate in a vacuum since no particles transmit the wave (Xiang et al., 2011).

### 2.11.2 How does Sound Wave Transfer Energy to Your Ears

Upon the arrival of sound waves at our auditory organs, they are directed towards the tympanic membrane through funneling. The tympanic membrane, commonly called the eardrum, is an integral component of the auditory system responsible for converting sound energy into mechanical vibrations. When subjected to impact, the object exhibits vibratory motion, which is transmitted through a fluid medium traversing three interconnected osseous structures. The fluid in motion causes the deflection of a sequence of cilia-like cells, which subsequently transform the vibrations into electrical signals within the nervous system. The transmission of these impulses to the brain occurs via the auditory nerves. The brain processes auditory stimuli and perceives them as sound (Grassi et al., 2013).

### 2.11.3 Measurement of Sound Energy

The unit of measurement for sound intensity is the decibel (dB), named after Alexander Graham Bell, a Scottish-born inventor. The audiometer, which measures an individual's auditory acuity, was developed by Bell. The decibel scale exhibits a logarithmic nature, indicating its ability to quantify the ratio between two observable physical quantities. Likewise, the measurement of sound pressure can also be denoted in decibels (dB). As an illustration, it can be observed that the sound level of an ordinary speaking voice typically registers at approximately 60 decibels. The acoustic level produced by the departure of an aircraft is measured at 115 decibels.

The unit of measurement for sound wave frequency is Hertz (Hz). A frequency of one Hertz corresponds to a single oscillation or vibration of a sound wave occurring within one second. For auditory perception to occur in humans, the frequency of the sound stimulus must fall within the range of 20 to 20,000 Hz. Infrasound refers to frequencies below 20 Hz, while ultrasound refers to frequencies above 20,000 Hz. The human ear is unable to perceive these two categories of sound (Tachibana et al., 1987).

### 2.11.4 Conversion of Sound Energy

According to the principle of energy conservation, energy is subject to the fundamental law that it cannot be generated or annihilated. The conversion of energy from one form to another is possible. Presented below are several illustrations of the conversion of sound energy into alternative forms and the reciprocal conversion of other forms into sound energy (Gambin et al., 2019).

- A microphone is capable of converting sound energy into electrical energy.
- A loudspeaker accomplishes the process of converting electricity into sound energy.
- The process by which an electric bell transforms electricity into sound energy is observed (Aghbashlo & Rosen, 2018).

### 2.11.5 Uses

- The principal application of sound energy pertains to the auditory perception process. In addition, it possesses various other applications.
- Infrasonic sound is utilized by scientists as a means of forecasting seismic events such as earthquakes, volcanic eruptions, and avalanches.
- The ocean floor is effectively mapped using a specific sound technology called SONAR (Sound Navigation and Ranging).
- Ultrasound is employed by medical professionals to therapeutically address tumor cells in cancer patients, effectively mitigating their growth and proliferation, all while ensuring the absence of any physical discomfort.
- Physicians also employ ultrasound technology to examine pregnant individuals to identify potential anomalies in the developing fetus.
- Ultrasound technology is employed within the industrial sector to identify potential defects in machinery components and assess the thickness of metallic materials.

- The utilization of Tibetan bowls emitting soothing musical sounds has been found to have the potential to alleviate stress and physical discomfort.
- Bats utilize ultrasonic waves for intra-species communication and obstacle detection during flight (Sakamoto et al., 2008).

## SUMMARY

Energy is the inherent ability to perform work or induce alterations, manifesting in diverse manifestations. Various forms of energy can be categorized according to their origin, inherent properties, and distinguishing attributes. Mechanical energy is intrinsically linked to an entity's dynamic or static state, either in terms of its movement or spatial arrangement. The concept can be further categorized into two distinct subtypes: kinetic and potential energy. Thermal energy refers to the energy linked to particles' internal motion within a given substance. Chemical energy storage occurs within the interatomic and intermolecular bonds present in a given substance. In the context of chemical reactions, energy is either released or absorbed. Electrical energy is the form of energy associated with the flow of electrons through conductive materials. Electromagnetic energy encompasses a wide range of energy, from radio waves and microwaves to visible light and gamma rays. Nuclear energy refers to the form of energy associated with the cohesive forces that bind the nucleus of atoms. Nuclear reactions, such as fission and fusion, release or absorb energy. Sound energy refers to the form of energy that arises from the oscillation of particles within a medium, commonly observed in the context of air. A comprehensive comprehension of these diverse manifestations of energy is imperative for many scientific and pragmatic implementations.

## REFERENCES

1. Aghbashlo, M., & Rosen, M. A. (2018). Exergoeconoenvironmental analysis as a new concept for developing thermodynamically, economically, and environmentally sound energy conversion systems. Journal of Cleaner Production, 187, 190-204.
2. Albanese, R., Blaschak, J., Medina, R., & Penn, J. (1994). Ultrashort electromagnetic signals: biophysical questions, safety issues, and medical opportunities. Aviation Space and Environmental Medicine, 65(5), A116.
3. Alva, G., Liu, L., Huang, X., & Fang, G. (2017). Thermal energy storage materials and systems for solar energy applications. Renewable and Sustainable Energy Reviews, 68, 693-706.
4. Anderson, J. S., Ayers, P. W., & Hernandez, J. I. R. (2010). How ambiguous is the local kinetic energy? The Journal of Physical Chemistry A, 114(33), 8884-8895.
5. Aslam, U., Rao, V. G., Chavez, S., & Linic, S. (2018). Catalytic conversion of solar to chemical energy on plasmonic metal nanostructures. Nature Catalysis, 1(9), 656-665.
6. Bailey, B. W., & McInnis, K. (2011). Energy cost of exergaming: a comparison of the energy cost of 6 forms of exergaming. Archives of pediatrics & adolescent medicine, 165(7), 597-602.
7. Baldocchi, D. (2014). Measuring fluxes of trace gases and energy between ecosystems and the atmosphere–the state and future of the eddy covariance method. Global change biology, 20(12), 3600-3609.
8. Ballardini, R., Balzani, V., Credi, A., Gandolfi, M. T., & Venturi, M. (2001). Artificial molecular-level machines: which energy to make them work?. Accounts of Chemical Research, 34(6), 445-455.
9. Baly, E. C. C. (1935). The kinetics of photosynthesis. Proceedings of the Royal Society of London. Series B-Biological Sciences, 117(804), 218-239.
10. Banerjee, S., Pillai, S. C., Falaras, P., O'Shea, K. E., Byrne, J. A., & Dionysiou, D. D. (2014). New insights into the mechanism of visible light photocatalysis. The journal of physical chemistry letters, 5(15), 2543-2554.
11. Batih, H., & Sorapipatana, C. (2016). Characteristics of urban households' electrical energy consumption in Indonesia and its saving potentials. Renewable and Sustainable Energy Reviews, 57, 1160-1173.
12. Bessman, S. P., & Geiger, P. J. (1981). Transport of energy in muscle: the phosphorylcreatine shuttle. Science, 211(4481), 448-452.

13. Biehs, S. A. (2007). Thermal heat radiation, near-field energy density and near-field radiative heat transfer of coated materials. The European Physical Journal B, 58, 423-431.
14. Biewener, A. A., & Roberts, T. J. (2000). Muscle and tendon contributions to force, work, and elastic energy savings: a comparative perspective. Exercise and Sport Sciences Reviews, 28(3), 99-107.
15. Boyer, D. (2011). Energopolitics and the Anthropology of Energy. Anthropology News, 52(5), 5-7.
16. Braams, B. J., & Bowman, J. M. (2009). Permutationally invariant potential energy surfaces in high dimensionality. International Reviews in Physical Chemistry, 28(4), 577-606.
17. Brevik, I. (1979). Experiments in phenomenological electrodynamics and the electromagnetic energy-momentum tensor. Physics Reports, 52(3), 133-201.
18. Brook, B. W., & Bradshaw, C. J. (2015). Key role for nuclear energy in global biodiversity conservation. Conservation Biology, 29(3), 702-712.
19. Bulthuis, D. A. (1987). Effects of temperature on photosynthesis and growth of seagrasses. Aquatic Botany, 27(1), 27-40.
20. Carminati, F., Roche, C., Rubio, J. A., Rubbia, C., Revol, J. P. C., & Klapisch, R. (1993). An energy amplifier for cleaner and inexhaustible nuclear energy production driven by a particle beam accelerator (93), 19698.
21. Celli, M., Colognesi, D., & Zoppi, M. (2000). Experimental determination of the translational kinetic energy of liquid and solid hydrogen. The European Physical Journal B-Condensed Matter and Complex Systems, 14, 239-244.
22. Chen, J. L., Li, G. S., & Wu, S. J. (2013). Assessing the potential of support vector machine for estimating daily solar radiation using sunshine duration. Energy conversion and management, 75, 311-318.
23. Cooper, M. M., & Klymkowsky, M. W. (2013). The trouble with chemical energy: Why understanding bond energies requires an interdisciplinary systems approach. CBE—Life Sciences Education, 12(2), 306-312.
24. Crawford, D. W. (1992). Metabolic cost of motility in planktonic protists: theoretical considerations on size scaling and swimming speed. Microbial Ecology, 24, 1-10.
25. Császár, A. G., & Handy, N. C. (1995). Exact quantum mechanical vibrational kinetic energy operator of sequentially bonded molecules in valence internal coordinates. The Journal of Chemical Physics, 102(10),

3962-3967.

26. De Groot, J. I., & Steg, L. (2010). Morality and nuclear energy: Perceptions of risks and benefits, personal norms, and willingness to take action related to nuclear energy. Risk Analysis: An International Journal, 30(9), 1363-1373.
27. De Strasser, L., Lipponen, A., Howells, M., Stec, S., & Bréthaut, C. (2016). A methodology to assess the water energy food ecosystems nexus in transboundary river basins. Water, 8(2), 59.
28. Demiroren, A., & Yilmaz, U. (2010). Analysis of change in electric energy cost with using renewable energy sources in Gökceada, Turkey: An island example. Renewable and Sustainable Energy Reviews, 14(1), 323-333.
29. Deser, S., & Tekin, B. (2002). Gravitational energy in quadratic-curvature gravities. Physical review letters, 89(10), 101101.
30. DeVany, A. S., Eckert, R. D., Meyers, C. J., & O'Hara, D. J. (1968). A property system for market allocation of the electromagnetic spectrum: A legal-economic-engineering study. Stan. L. Rev., 21, 1499.
31. Diddams, S. A., Vahala, K., & Udem, T. (2020). Optical frequency combs: Coherently uniting the electromagnetic spectrum. Science, 369(6501), 3676.
32. Dunn, B., Kamath, H., & Tarascon, J. M. (2011). Electrical energy storage for the grid: a battery of choices. Science, 334(6058), 928-935.
33. Dupont, E., Koppelaar, R., & Jeanmart, H. (2020). Global available solar energy under physical and energy return on investment constraints. Applied Energy, 257, 113968.
34. Emerson, R., & Arnold, W. (1932). The photochemical reaction in photosynthesis. The Journal of General Physiology, 16(2), 191-205.
35. Evans, O., Caragata, E. P., McMeniman, C. J., Woolfit, M., Green, D. C., Williams, C. R., ... & McGraw, E. A. (2009). Increased locomotor activity and metabolism of Aedes aegypti infected with a life-shortening strain of Wolbachia pipientis. Journal of Experimental Biology, 212(10), 1436-1441.
36. Falk, G., Herrmann, F., & Schmid, G. B. (1983). Energy forms or energy carriers?. American Journal of Physics, 51(12), 1074-1077.
37. Fenchel, T., & Finlay, B. J. (1983). Respiration rates in heterotrophic, free-living protozoa. Microbial Ecology, 9, 99-122.
38. Ferrari, R., & Wunsch, C. (2009). Ocean circulation kinetic energy: Reservoirs, sources, and sinks. Annual Review of Fluid Mechanics, 41, 253-282.

39. Ferreira, H. L., Garde, R., Fulli, G., Kling, W., & Lopes, J. P. (2013). Characterisation of electrical energy storage technologies. Energy, 53, 288-298.

40. Fjørtoft, R. (1953). On the changes in the spectral distribution of kinetic energy for two dimensional, nondivergent flow. Tellus, 5(3), 225-230.

41. Fleming, G. R., Martin, J. L., & Breton, J. (1988). Rates of primary electron transfer in photosynthetic reaction centres and their mechanistic implications. Nature, 333(6169), 190-192.

42. Flesch, L. M., Holt, W. E., Haines, A. J., Wen, L., & Shen-Tu, B. (2007). The dynamics of western North America: Stress magnitudes and the relative role of gravitational potential energy, plate interaction at the boundary and basal tractions. Geophysical Journal International, 169(3), 866-896.

43. Gallo, M., Mescia, L., Losito, O., Bozzetti, M., & Prudenzano, F. (2012). Design of optical antenna for solar energy collection. Energy, 39(1), 27-32.

44. Gambin, B., Kruglenko, E., Tymkiewicz, R., & Litniewski, J. (2019). Ultrasound assessment of the conversion of sound energy into heat in tissue phantoms enriched with magnetic micro-and nanoparticles. Medical Physics, 46(10), 4361-4370.

45. Garten Jr, C. T. (1978). A review of parameter values used to assess the transport of plutonium, uranium, and thorium in terrestrial food chains. Environmental Research, 17(3), 437-452.

46. Ghosh, A., Holt, W. E., Flesch, L. M., & Haines, A. J. (2006). Gravitational potential energy of the Tibetan Plateau and the forces driving the Indian plate. Geology, 34(5), 321-324.

47. Grassi, M., Pastore, M., & Lemaitre, G. (2013). Looking at the world with your ears: How do we get the size of an object from its sound?. Acta Psychologica, 143(1), 96-104.

48. Gupta, C. K., & Krishnamurthy, N. (1992). Extractive metallurgy of rare earths. International materials reviews, 37(1), 197-248.

49. Gutowski, T., Dahmus, J., & Thiriez, A. (2006, May). Electrical energy requirements for manufacturing processes. In 13th CIRP international conference on life cycle engineering. Leuven, Belgium. (Vol. 31, pp. 623-638).

50. Hamada, N., & Ogino, H. (2012). Food safety regulations: what we learned from the Fukushima nuclear accident. Journal of environmental radioactivity, 111, 83-99.

51. Hameer, S., & van Niekerk, J. L. (2015). A review of large-scale electrical energy storage. International Journal of Energy Research, 39(9), 1179-1195.
52. Hannon, B. (1973). The structure of ecosystems. Journal of Theoretical Biology, 41(3), 535-546.
53. Hastik, R., Basso, S., Geitner, C., Haida, C., Poljanec, A., Portaccio, A., ... & Walzer, C. (2015). Renewable energies and ecosystem service impacts. Renewable and Sustainable Energy Reviews, 48, 608-623.
54. Hatzfeld, D., Martinod, J., Bastet, G., & Gautier, P. (1997). An analog experiment for the Aegean to describe the contribution of gravitational potential energy. Journal of Geophysical Research: Solid Earth, 102(B1), 649-659.
55. Hayward, S. A. (1996). Gravitational energy in spherical symmetry. Physical Review D, 53(4), 1938.
56. Hill, D. J. (2008). Nuclear energy for the future. Nature Materials, 7(9), 680-682.
57. Horowitz, G. T., & Perry, M. J. (1982). Gravitational energy cannot become negative. Physical Review Letters, 48(6), 371.
58. Hsiao, K. L. (2017). To promote radiation electrical MHD activation energy thermal extrusion manufacturing system efficiency by using Carreau-Nanofluid with parameters control method. Energy, 130, 486-499.
59. Hubbert, M. K. (1956). Nuclear energy and the fossil fuels. Houston, TX: Shell Development Company, Exploration and Production Research Division. (Vol. 95, pp. 74-78).
60. Jaupart, C., Labrosse, S., Lucazeau, F., & Mareschal, J. C. (2007). 7.06-temperatures, heat and energy in the mantle of the earth. Treatise on geophysics, 7, 223-270.
61. Jennings, L. D., Miller, R. E., & Spedding, F. H. (1960). Lattice heat capacity of the rare earths. Heat capacities of yttrium and lutetium from 15–350 k. The Journal of Chemical Physics, 33(6), 1849-1852.
62. Johri, V. B., Kalligas, D., Singh, G. P., & Everitt, C. W. F. (1995). Gravitational energy in the expanding universe. General Relativity and Gravitation, 27, 313-318.
63. Jones, C. H., Unruh, J. R., & Sonder, L. J. (1996). The role of gravitational potential energy in active deformation in the southwestern United States. Nature, 381(6577), 37-41.
64. Joulain, K., Mulet, J. P., Marquier, F., Carminati, R., & Greffet, J. J.

(2005). Surface electromagnetic waves thermally excited: Radiative heat transfer, coherence properties and Casimir forces revisited in the near field. Surface Science Reports, 57(3-4), 59-112.

65. Jun, E., Kim, W. J., Jeong, Y. H., & Chang, S. H. (2010). Measuring the social value of nuclear energy using contingent valuation methodology. Energy Policy, 38(3), 1470-1476.
66. Kaper, W. H., & Goedhart, M. J. (2002). 'Forms of energy', an intermediary language on the road to thermodynamics? Part II. International Journal of Science Education, 24(2), 119-137.
67. Kay, S., Hedley, J. D., & Lavender, S. (2009). Sun glint correction of high and low spatial resolution images of aquatic scenes: A review of methods for visible and near-infrared wavelengths. Remote sensing, 1(4), 697-730.
68. Kemp, W., & Kemp, W. (1991). Energy and the electromagnetic spectrum. Organic spectroscopy, 4(3), 1-17.
69. Khaligh, A., Zeng, P., & Zheng, C. (2009). Kinetic energy harvesting using piezoelectric and electromagnetic technologies—state of the art. IEEE transactions on industrial electronics, 57(3), 850-860.
70. Khan, N., Dilshad, S., Khalid, R., Kalair, A. R., & Abas, N. (2019). Review of energy storage and transportation of energy. Energy Storage, 1(3), 49.
71. Klots, C. E. (1976). Kinetic energy distributions from unimolecular decay: Predictions of the Langevin model. The Journal of Chemical Physics, 64(11), 4269-4275.
72. Kondoh, J., Ishii, I., Yamaguchi, H., Murata, A., Otani, K., Sakuta, K., ... & Kamimoto, M. (2000). Electrical energy storage systems for energy networks. Energy conversion and management, 41(17), 1863-1874.
73. Koning, A. J., & Rochman, D. (2008). Towards sustainable nuclear energy: Putting nuclear physics to work. Annals of Nuclear Energy, 35(11), 2024-2030.
74. Kronig, R. D. L. (1926). On the theory of dispersion of x-rays. Josa, 12(6), 547-557.
75. Kumarasamy, K., An, J., Yang, J., & Yang, E. H. (2017). Novel CFD-based numerical schemes for conduction dominant encapsulated phase change materials (EPCM) with temperature hysteresis for thermal energy storage applications. Energy, 132, 31-40.
76. Lenzen, M. (2008). Life cycle energy and greenhouse gas emissions of nuclear energy: A review. Energy conversion and management, 49(8), 2178-2199.
77. Lewis, N. S., Crabtree, G., Nozik, A. J., Wasielewski, M. R., Alivisatos, P.,

Kung, H., ... & Nault, R. M. (2005). Basic research needs for solar energy utilization. Report of the basic energy sciences workshop on solar energy utilization, April 18-21, 2005. DOESC (USDOE Office of Science (SC)). 41(3), 987-988.

78. Lindsey, B. A. (2014). Student reasoning about electrostatic and gravitational potential energy: An exploratory study with interdisciplinary consequences. Physical Review Special Topics-Physics Education Research, 10(1), 013101.
79. Liu, G., Leng, Q., Lian, J., Guo, H., Yi, X., & Hu, C. (2015). Notepad-like triboelectric generator for efficiently harvesting low-velocity motion energy by interconversion between kinetic energy and elastic potential energy. ACS applied materials & interfaces, 7(2), 1275-1283.
80. Magdziarz, P., & Zdziarski, A. A. (1995). Angle-dependent Compton reflection of X-rays and gamma-rays. Monthly Notices of the Royal Astronomical Society, 273(3), 837-848.
81. Mahto, R. P., Rout, M., & Pal, S. K. (2021). Mechanism of microstructure evolution and grain growth in friction stir welding of AA6061-T6 and AISI304 in air and water media. Materials Chemistry and Physics, 273, 125081.
82. Maier, S. A., & Atwater, H. A. (2005). Plasmonics: Localization and guiding of electromagnetic energy in metal/dielectric structures. Journal of Applied Physics, 98(1), 8741-8742.
83. Makarieva, A. M., Gorshkov, V. G., Li, B. L., Chown, S. L., Reich, P. B., & Gavrilov, V. M. (2008). Mean mass-specific metabolic rates are strikingly similar across life's major domains: evidence for life's metabolic optimum. Proceedings of the National Academy of Sciences, 105(44), 16994-16999.
84. Maluf, J. W. (2005). The gravitational energy-momentum tensor and the gravitational pressure. Annalen der Physik, 517(11-12), 723-732.
85. Maroti, P., & Wraight, C. A. (1997). Kinetics of H+ ion binding by the P+ QA-state of bacterial photosynthetic reaction centers: rate limitation within the protein. Biophysical Journal, 73(1), 367-381.
86. Marshall, L. G. (1994). An acoustics measurement program for evaluating auditoriums based on the early/late sound energy ratio. The Journal of the Acoustical Society of America, 96(4), 2251-2261.
87. McKendry, P. (2002). Energy production from biomass (part 1): Overview of biomass. Bioresource Technology, 83(1), 37-46.
88. Meng, X., Liu, L., Ouyang, S., Xu, H., Wang, D., Zhao, N., & Ye, J. (2016). Nanometals for solar-to-chemical energy conversion: from

semiconductor-based photocatalysis to plasmon-mediated photocatalysis and photo-thermocatalysis. Advanced Materials, 28(32), 6781-6803.

89. Moeller, T., & Ferrus, R. (1961). Observations on the rare earths—LXXIII the heat and entropy of formation of the 1: 1 chelates of N-hydroxyet hylethylenediaminetriacetic acid with the tripositive cations. Journal of Inorganic and Nuclear Chemistry, 20(3-4), 261-273.

90. Myers, J. A. C. K. (1971). Enhancement studies in photosynthesis. Annual Review of Plant Physiology, 22(1), 289-312.

91. Ngamroo, I., & Vachirasricirikul, S. (2013). Optimized SFCL and SMES units for multimachine transient stabilization based on kinetic energy control. IEEE transactions on applied superconductivity, 23(3), 5000309-5000309.

92. Nia, H. T., Liu, H., Seano, G., Datta, M., Jones, D., Rahbari, N., ... & Jain, R. K. (2016). Solid stress and elastic energy as measures of tumour mechanopathology. Nature Biomedical Engineering, 1(1), 0004.

93. Occhiello, E., Morra, M., Morini, G., Garbassi, F., & Humphrey, P. (1991). Oxygen-plasma-treated polypropylene interfaces with air, water, and epoxy resins: Part I. Air and water. Journal of Applied Polymer Science, 42(2), 551-559.

94. Odum, E. P. (1968). Energy flow in ecosystems: a historical review. American Zoologist, 8(1), 11-18.

95. Osepchuk, J. M., & Petersen, R. C. (2003). Historical review of RF exposure standards and the International Committee on Electromagnetic Safety (ICES). Bioelectromagnetics, 24(S6), S7-S16.

96. Perera, W. P. T. D., Navaratne, S. B., & Wickramasinghe, I. (2022). Review on effect of postharvest illumination by fluorescent and ultraviolet light waves on the quality of vegetables. Journal of Food Process Engineering, 45(2), e13960.

97. Petrescu, F. I., Apicella, A., Petrescu, R. V., Kozaitis, S., Bucinell, R., Aversa, R., & Abu-Lebdeh, T. (2016). Environmental protection through nuclear energy. American Journal of Applied Sciences, 13(9), 941-946.

98. Picchi, P., van Lierop, M., Geneletti, D., & Stremke, S. (2019). Advancing the relationship between renewable energy and ecosystem services for landscape planning and design: A literature review. Ecosystem Services, 35, 241-259.

99. Porter, J., Costanza, R., Sandhu, H., Sigsgaard, L., & Wratten, S. (2009). The value of producing food, energy, and ecosystem services within an agro-ecosystem. AMBIO: A Journal of the Human Environment, 38(4), 186-193.

100. Prommas, R., Rattanadecho, P., & Cholaseuk, D. (2010). Energy and exergy analyses in drying process of porous media using hot air. International Communications in Heat and Mass Transfer, 37(4), 372-378.

101. Psaltis, D. (2008). Probes and tests of strong-field gravity with observations in the electromagnetic spectrum. Living Reviews in Relativity, 11, 1-61.

102. Quinten, M., Leitner, A., Krenn, J. R., & Aussenegg, F. R. (1998). Electromagnetic energy transport via linear chains of silver nanoparticles. Optics letters, 23(17), 1331-1333.

103. Raina, J. B., Giardina, M., Brumley, D. R., Clode, P. L., Pernice, M., Guagliardo, P., ... & Seymour, J. R. (2023). Chemotaxis increases metabolic exchanges between marine picophytoplankton and heterotrophic bacteria. Nature Microbiology, 8(3), 510-521.

104. Ranius, T., Hämäläinen, A., Egnell, G., Olsson, B., Eklöf, K., Stendahl, J., ... & Felton, A. (2018). The effects of logging residue extraction for energy on ecosystem services and biodiversity: A synthesis. Journal of environmental management, 209, 409-425. Ranius, T., Hämäläinen, A., Egnell, G., Olsson, B., Eklöf, K., Stendahl, J., ... & Felton, A. (2018). The effects of logging residue extraction for energy on ecosystem services and biodiversity: A synthesis. Journal of Environmental Management, 209, 409-425.

105. Roeser, S. (2011). Nuclear energy, risk, and emotions. Philosophy & Technology, 24, 197-201.

106. Rosen, A., Stuchly, M. A., & Vander Vorst, A. (2002). Applications of RF/microwaves in medicine. IEEE transactions on microwave theory and techniques, 50(3), 963-974.

107. Routley, R., & Routley, V. (1978). Nuclear energy and obligations to the future. Inquiry, 21(1-4), 133-179.

108. Sakamoto, S. I., Nishikawa, M., Ishino, T., Watanabe, Y., & Senda, J. (2008). Effect of inner diameter change of phase adjuster on heat-to-sound energy conversion efficiency in loop-tube-type thermoacoustic prime mover. Japanese journal of applied physics, 47(5S), 4223.

109. Sato, S. (1955). A new method of drawing the potential energy surface. Bulletin of the Chemical Society of Japan, 28(7), 450-453.

110. Sayvetz, A. (1939). The kinetic energy of polyatomic molecules. The Journal of Chemical Physics, 7(6), 383-389.

111. Schlegel, H. B. (2003). Exploring potential energy surfaces for chemical reactions: an overview of some practical methods. Journal of Computational Chemistry, 24(12), 1514-1527.

112. Seddegh, S., Joybari, M. M., Wang, X., & Haghighat, F. (2017). Experimental and numerical characterization of natural convection in a vertical shell-and-tube latent thermal energy storage system. Sustainable Cities and Society, 35, 13-24.
113. Sharma, A., Tyagi, V. V., Chen, C. R., & Buddhi, D. (2009). Review on thermal energy storage with phase change materials and applications. Renewable and Sustainable Energy Reviews, 13(2), 318-345.
114. Shi, X., Meng, Y., Bi, R., Wan, Z., Zhu, Y., & Rojas, O. J. (2022). Enabling unidirectional thermal conduction of wood-supported phase change material for photo-to-thermal energy conversion and heat regulation. Composites Part B: Engineering, 245, 110231.
115. Silverstein, D. L., & Fort, T. (2000). Incorporating Low Hydraulic Conductivity in a Numerical Model for Predicting Air– Water Interfacial Area in Wet Unsaturated Particulate Porous Media. Langmuir, 16(2), 835-838.
116. Sonnberger, M., Ruddat, M., Arnold, A., Scheer, D., Poortinga, W., Böhm, G., ... & Tvinnereim, E. (2021). Climate concerned but anti-nuclear: Exploring (dis) approval of nuclear energy in four European countries. Energy Research & Social Science, 75, 102008.
117. Spedding, F. H., & Henderson, D. C. (1971). High-Temperature Heat Contents and Related Thermodynamic Functions of Seven Trifluorides of the Rare Earths: Y, La, Pr, Nd, Gd, Ho, and Lu. The Journal of Chemical Physics, 54(6), 2476-2483.
118. Stein, C., Lowman, J. P., & Hansen, U. (2013). The influence of mantle internal heating on lithospheric mobility: Implications for super-Earths. Earth and Planetary Science Letters, 361, 448-459.
119. Stiner, M. C., & Feeley-Harnik, G. (2011). Energy and ecosystems. Deep history: The architecture of past and present, 10, 78-102.
120. Straede, C., Budtz-Jørgensen, C., & Knitter, H. H. (1987). 235U (n, f) Fragment mass-, kinetic energy and angular distributions for incident neutron energies between thermal and 6 MeV. Nuclear Physics A, 462(1), 85-108.
121. Strumpen, N. F., Rädecker, N., Pogoreutz, C., Meibom, A., & Voolstra, C. R. (2022). High light quantity suppresses locomotion in symbiotic Aiptasia. Symbiosis, 86(3), 293-304.
122. Tachibana, H., Yano, H., & Yoshihisa, K. (1987). Definition and measurement of sound energy level of a transient sound source. Journal of the Acoustical Society of Japan (E), 8(6), 235-240.

123. Tal, Y., & Bader, R. F. W. (1978). Studies of the energy density functional approach. I. Kinetic energy. International Journal of Quantum Chemistry, 14(S12), 153-168.

124. Tan, H., Hirst, A. G., Glazier, D. S., & Atkinson, D. (2019). Ecological pressures and the contrasting scaling of metabolism and body shape in coexisting taxa: cephalopods versus teleost fish. Philosophical Transactions of the Royal Society B, 374(1778), 20180543.

125. Tang, A., Xu, Y., Pan, J., Yang, W., Jiang, D., & Lu, Q. (2015). Combustion characteristics and performance evaluation of premixed methane/air with hydrogen addition in a micro-planar combustor. Chemical Engineering Science, 131, 235-242.

126. Thayer, J. P., Vickrey, J. F., Heelis, R. A., & Gary, J. B. (1995). Interpretation and modeling of the high-latitude electromagnetic energy flux. Journal of Geophysical Research: Space Physics, 100(10), 19715-19728.

127. Thomas, P. A., & Gates, T. E. (1999). Radionuclides in the lichen-caribou-human food chain near uranium mining operations in northern Saskatchewan, Canada. Environmental Health Perspectives, 107(7), 527-537.

128. Tian, Y., & Zhao, C. Y. (2013). A review of solar collectors and thermal energy storage in solar thermal applications. Applied Energy, 104, 538-553.

129. Tolmachev, Y. V., Piatkivskyi, A., Ryzhov, V. V., Konev, D. V., & Vorotyntsev, M. A. (2015). Energy cycle based on a high specific energy aqueous flow battery and its potential use for fully electric vehicles and for direct solar-to-chemical energy conversion. Journal of Solid State Electrochemistry, 19, 2711-2722.

130. Valencia, D., & O'Connell, R. J. (2009). Convection scaling and subduction on Earth and super-Earths. Earth and Planetary Science Letters, 286(3-4), 492-502.

131. Van de Hulst, H. C. (1945). Radiogolven uit het wereldruim: II. Herkomst der radiogolven; Radiogolven uit het wereldruim: II. Herkomst der radiogolven; Radio waves from space. Nederlandsch Tijdschrift voor Natuurkunde, 11, 210-221.

132. Van der Pligt, J., Eiser, J. R., & Spears, R. (1986). Attitudes toward nuclear energy: Familiarity and salience. Environment and Behavior, 18(1), 75-93.

133. Van Heck, H. J., & Tackley, P. J. (2011). Plate tectonics on super-Earths: equally or more likely than on Earth. Earth and Planetary Science Letters, 310(3-4), 252-261.

134. Vogel, J., & Thess, A. (2019). Validation of a numerical model with a benchmark experiment for melting governed by natural convection in latent thermal energy storage. Applied Thermal Engineering, 148, 147-159.

135. Walker, G. (2008). What are the barriers and incentives for community-owned means of energy production and use? Energy Policy, 36(12), 4401-4405.

136. Wang, L. W., & Teter, M. P. (1992). Kinetic-energy functional of the electron density. Physical Review B, 45(23), 13196.

137. Wasielewski, M. R. (1992). Photoinduced electron transfer in supramolecular systems for artificial photosynthesis. Chemical Reviews, 92(3), 435-461.

138. Watts, D. M. (1983). Some alternative views of energy. Physics education, 18(5), 213.

139. White, L. J. (2001). "Propertyzing" The Electromagnetic Spectrum: Why It's Important, And How To Begin. In Communications deregulation and FCC reform: Finishing the job. Boston, MA: Springer US. (Vol. 3, pp. 111-143).

140. Whittingham, M. S. (2008). Materials challenges facing electrical energy storage. Mrs Bulletin, 33(4), 411-419.

141. Wilson, J. M., & Flanagan, E. P. (2008). The role of elastic energy in activities with high force and power requirements: a brief review. The Journal of Strength & Conditioning Research, 22(5), 1705-1715.

142. Xiang, N., Goggans, P., Jasa, T., & Robinson, P. (2011). Bayesian characterization of multiple-slope sound energy decays in coupled-volume systems. The Journal of the Acoustical Society of America, 129(2), 741-752.

143. Yamada, N. (2015). Kinetic energy discrimination in collision/reaction cell ICP-MS: Theoretical review of principles and limitations. Spectrochimica Acta Part B: Atomic Spectroscopy, 110, 31-44.

144. Yu, M., Han, Y., Cheng, X., Hu, L., Zeng, Y., Chen, M., ... & Tong, Y. (2015). Holey tungsten oxynitride nanowires: novel anodes efficiently integrate microbial chemical energy conversion and electrochemical energy storage. Advanced materials, 27(19), 3085-3091.

145. Zeng, H., Ellis, C. S., & Lebeck, A. R. (2005). Experiences in managing energy with ecosystem. IEEE Pervasive Computing, 4(1), 62-68.

146. Zhou, L., Yang, T., Fang, Z., Zhou, J., Zheng, Y., Guo, C., ... & Wang, Z. L. (2022). Boosting of water splitting using the chemical energy

simultaneously harvested from light, kinetic energy and electrical energy using N doped 4H-SiC nanohole arrays. Nano Energy, 104, 107876.

147. Zinkle, S. J., & Was, G. S. (2013). Materials challenges in nuclear energy. Acta Materialia, 61(3), 735-758.

CHAPTER

# 3 Biological and Chemical Energy

## Contents

## INTRODUCTION

Energy is a fundamental requirement for the sustenance of living organisms and the various processes carried out by living cells. The sustenance of life on Earth is contingent upon the accessibility of solar energy. Plants utilize the process of photosynthesis to convert sunlight into chemical energy, that is subsequently employed to sustain their metabolic functions and facilitate growth. The energy that is acquired by plants is utilized by all other living organisms, either directly or indirectly (Golueke & Oswald, 1959).

An individual devoid of physical exertion requires a total energy intake of 1700 kilocalories over 24 hours to sustain vital physiological functions. The aforementioned quantity of energy can be obtained through the intake and metabolism of 200 grams (equivalent to 800 kilocalories) of carbohydrates, 70 grams (equivalent to 400 kilocalories) of protein, and 55 grams (equivalent to 500 kilocalories) of fat. Furthermore, cellular respiration results in the consumption of 16 moles of oxygen and the production of 13.7 moles of carbon dioxide. Individuals who engage in regular physical activity require a daily caloric intake ranging from 3000 to 3500 kcal. However, those who engage in physically demanding occupations or activities may have an increased caloric requirement of up to 5000 kcal (Humphris et al., 1995).

The utilization of biological sources of energy directly contributes to a substantial proportion of the overall energy consumption by the human population. In addition to its utilization as a source of sustenance, heat, and culinary purposes, biomass derived from a diverse range of sources is employed for the generation of power. Sugarcane as well as maize are subjected to a conversion process that results in the production of ethanol and various other biofuels (Qian, 2007).

Fossil fuels, namely coal, petroleum, as well as natural gas, are derived from organic energy sources. In contrast to biological energy, the organic material of interest underwent natural processing within extended temporal intervals.

This chapter examines energy resources that have undergone a conversion process from solar energy through the mechanism of photosynthesis. Our primary focus is directed towards the underlying processes of photosynthesis. Next, we conduct a comprehensive survey comparing biofuels as well as fossil fuels (Kamerlin & Warshel, 2010).

## 3.1 PHOTOSYNTHESIS

Photosynthesis is a fundamental biochemical process employed by the majority of plants, algae, and cyanobacteria, enabling them to harness and convert solar energy into chemical energy for storage purposes. The chemical energy is stored within carbohydrate molecules, specifically sugars, that are formed through the synthesis of carbon dioxide and water (Ke, 2001).

The central participant in the process of photosynthesis in the majority of plants is the chlorophyll molecule. The green coloration observed in numerous plants and algae can be attributed to the presence of chlorophyll. The concentration of chlorophyll occurs intracellularly within specialized organelles known as chloroplasts. Plants are commonly observed to possess a green coloration due to the predominant absorption of blue and red wavelengths by chlorophyll, resulting in the reflection of green light. Multiple variants of chlorophyll exist, albeit possessing a common chlorin-magnesium ligand. Chlorophyll performs three primary roles within the context of photosynthesis: (i) light absorption, (ii) the transfer of resonance energy to a designated chlorophyll pair located in the center of the reaction of the photosystems, and (iii) energy transfer to an electron through a process known as charge separation, ultimately facilitating biosynthesis (Evans, 2013).

The process of photosynthesis in physics exhibits numerous parallels with the energy-gathering mechanism observed in a photovoltaic (PV) cell. Photons that are absorbed can excite electrons in a manner that is comparable to the stimulation of electrons throughout the band gap of silicon within a photovoltaic solar cell. The energy required for photosynthesis is derived from the excited electrons. Nevertheless, within biological systems, the electrochemical energy acquired through the process of photosynthesis accumulates in organic compounds predominantly consisting of carbohydrates, including sugars and starches (Farquhar et al., 2001).

From an energetic perspective, carbohydrates are considered to be the most crucial organic molecules. These molecules are classified as relatively uncomplicated compounds composed of carbon, oxygen, and hydrogen, which serve the purpose of storing and facilitating the transfer of energy within biological systems. The majority of carbohydrates can be represented by the chemical formula C(n)⟦(H2O)⟧m, which is derived from the structural composition of carbon and water molecules. This nomenclature, "carbo-hydrate," reflects the bond arrangement within these compounds. The compounds can be classified into four distinct chemical categories, namely monosaccharides, disaccharides, oligosaccharides, as well as polysaccharides. The initial two substances are commonly denoted as sugars and encompass fructose (fruit sugar), sucrose (cane or beet sugar), lactose (milk sugar), and glucose (starch sugar). Complex carbohydrates such as cellulose, hemicellulose, and lignin are constituents utilized by plants to provide structural support (Eberhard et al., 2008).

### 3.1.1 Mechanisms of Photosynthesis

There exist two distinct categories of photosynthetic processes: oxygenic photosynthesis as well as oxygenic photosynthesis. Both of these processes exhibit similarities, however, it is worth noting that oxygenic photosynthesis is the prevailing mechanism and is observed in various organisms such as plants, algae, and cyanobacteria (Raven et al., 2008).

In the process of oxygenic photosynthesis, the transfer of light energy facilitates the movement of electrons from water to carbon dioxide, ultimately resulting in the synthesis of carbohydrates. During this process, carbon dioxide undergoes reduction by gaining electrons, while water undergoes oxidation by losing electrons. Oxygenic photosynthesis serves as a mechanism to offset respiration by absorbing the carbon dioxide emitted by all living organisms and subsequently releasing oxygen back into the atmosphere (González et al., 2021).

The intricacies of the processes underlying photosynthesis are highly intricate, surpassing the boundaries of this book's scope for a comprehensive elucidation. Therefore, our attention is directed solely toward a specific process, while disregarding the more intricate mechanisms that are implicated. Figure 3.1 depicts a simplified illustration of oxygenic photosynthesis, showcasing the sequential reactions involved in the fixation of carbon dioxide by certain plant species. The aforementioned process is commonly referred to as the Calvin-Benson-Bassham (CBB) cycle, alternatively known as the Calvin cycle. The aforementioned cycle encompasses a total of thirteen biochemical reactions, all of which are facilitated by eleven unique molecules. The overall outcome entails the transformation of three molecules of carbon dioxide ($CO_2$) into a sugar known as triose-P, which consists of three carbon atoms. This process is facilitated by utilizing the energy stored within adenosine triphosphate (ATP) as well as nicotinamide adenine dinucleotide phosphate (NADPH), which are molecules involved in energy transport. The energy in question is derived from the light reactions of photosynthesis. Triose-P is subsequently utilized in the synthesis of hexose sugars and sugar phosphates, which serve as reservoirs for energy storage in the form of stable compounds like glucose or sucrose. The enzyme RuBisCO, which stands for Ribulose-1, 5-bisphosphate carboxylase/oxygenase, plays a pivotal role in facilitating the carboxylation of ribulose-1, 5-bisphosphate (RuBP). This process involves the capture of carbon dioxide ($CO_2$) from the atmosphere (Clayton, 1972).

In contrast to oxygenic photosynthesis, anoxygenic photosynthesis employs electron donors that are distinct from water. The process commonly takes place in bacterial organisms, specifically purple bacteria as well as green sulfur bacteria, predominantly inhabiting diverse aquatic environments.

Photosynthesis encompasses a complex series of biochemical reactions that result in the following net conversion (Bach et al., 2013):

$$n(8\gamma + CO_2 + H_2O) \rightarrow C_nH_{2n}O_n + nO_2 \qquad (3.1)$$

The observed reaction demonstrates that when subjected to the influence of 8n photons, a combination of n-water as well as carbon dioxide molecules yields a single molecule of sugar as well as n molecules of oxygen. The sugars generated through the process of photosynthesis can be utilized in the synthesis of starches and cellulose, serving as a means of energy storage and contributing to the structural development of plants (Liu et al., 2019).

### 3.1.2 Photosynthesis Efficiency

The conversion efficiency of solar energy into biomass for the majority of plant species is typically less than 1%. The photosynthetically active frequency spectrum encompasses only 40% of the total energy present in sunlight.

The process of capturing photons involves energy loss during the transportation of electrons to the reaction site, as well as at each stage of the chemical reaction. Consequently, the proportion of energy stored in an individual $CH_2O$ sugar component, derived from the absorption of eight photons at a wavelength of 680 nm, is approximately 30%. The upper conceptual limit of photosynthetic efficiency is estimated to be 12% (Simkin et al., 2019).

The diagram presented in Figure 3.1 is A simplified diagram illustrating the C-3 photosynthesis process, specifically the binding of carbon dioxide (CO2) in the Calvin cycle, which occurs in two distinct stages. The process of photosynthesis can be divided into two main stages: (1) light reactions, which involve the absorption of light energy to produce ATP and NADPH molecules, and (2) light-independent reactions, which utilize these products to capture while lowering carbon dioxide.

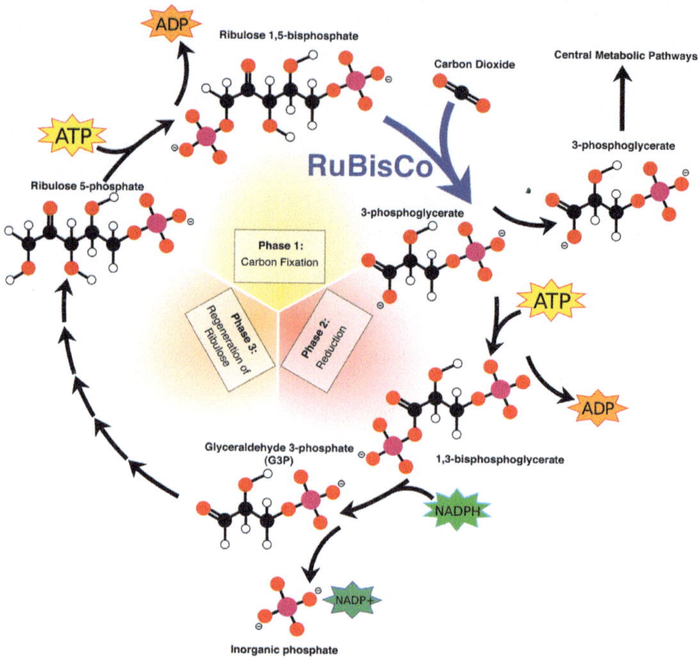


**Figure 3.1.** A simplified schematic representation of the C-3 photosynthesis process, specifically the CO2 binding step in the Calvin cycle, which takes place in two distinct stages. There are two distinct processes involved in photosynthesis. The first process, known as light reactions, involves the absorption of light energy to generate ATP and NADPH molecules. The second process, referred to as light-independent reactions, utilizes these aforementioned products to capture and subsequently reduce carbon dioxide. (Source: By Mike Jones - Own work, CC BY-SA 3.0, https://commons.wikimedia.org/w/index.php?curid=9504547 , Creative Commons License).

It is important to note that ATP stands for adenosine triphosphate, NADPH stands for reduced nicotinamide adenine dinucleotide phosphate, NADP+ stands for nicotinamide adenine dinucleotide phosphate, ADP stands for adenosine diphosphate, and Pi represents inorganic phosphate. The achievable maximum efficiency is approximately 5%, while the prevailing efficiencies tend to be significantly lower. This phenomenon can be attributed to the fact that a significant portion of the energy that transforms sugars is rapidly utilized for the metabolism of plants and various other functions (Hu et al., 2013). The amount of energy contained within the plant's structural material, which can be harnessed through combustion or other methods, is generally significantly lower than 1% of the incoming sunlight, typically averaging around 0.25%. The global rate of production of biomass on Earth, which is estimated to be 1660 EJ/y, serves as a demonstration of the relatively low efficiency of photosynthesis. This is particularly evident when considering the rate of energy consumption by humankind, which stands at 550 EJ/y (Emerson & Lewis, 1942).

## 3.2 FOOD ENERGY

The annual rate of generation of energy for consumption by humans is estimated to be approximately 45 exajoules (EJ), with approximately one-third of this energy being wasted (Harvey & Pilgrim, 2011).

### 3.2.1 Food Production

Approximately 50 million square kilometers (equivalent to approximately one-third of the total land area of the world) are allocated for agricultural purposes. Out of this, 15 million square kilometers are specifically designated for crop cultivation, while the remaining land is utilized as pastures and grasslands for grazing livestock. Maize, rice, and wheat are widely recognized as the three primary crops, collectively accounting for approximately 60% of the global food supply. Considering the land area allocated for cultivating the aforementioned trio of crops and their respective yields, it can be inferred that the efficiency pertaining to the capture of solar energy is approximately 0.1% (Caplice & Fitzgerald, 1999).

### 3.2.2 Fertilizers

Plants necessitate nitrogen for their growth, as nitrogen is integrated into the composition of certain vital complex molecules, such as DNA and amino acids, which are essential for plant development. The scarcity of nitrogen typically serves as the primary constraint on plant growth in the majority of natural environments. The composition of the atmosphere includes nitrogen primarily in the form of N2, a diatomic molecule characterized by a strong triple bond that renders it highly resistant to dissociation. Specific microorganisms, namely cyanobacteria, play a vital role in the nitrogen cycle by actively engaging in the process of nitrogen fixation. This process involves the conversion of atmospheric nitrogen into ammonia through the following chemical reaction: $N_2$

$$N_2 + 8H^+ + 8e^- \rightarrow 2NH_3 + H_2$$

Nevertheless, the cultivation of high-yield plants necessitates the utilization of synthetic ammonia, which is generated through the energy-intensive Haber-Bosch process, consuming approximately 36 MJ/kg. The utilization of fixed nitrogen in fertilizers is believed to contribute to approximately 50% of the food energy consumed. Nevertheless, the practice of intensive agriculture as well as monoculture has been found to result in a decline in soil quality and an escalation in pest populations. The necessity for heightened pesticide usage contributes to the exacerbation of environmental issues (Souri, 2016).

## 3.3 BIOENERGY

Bioenergy refers to the form of energy derived from organic matter. Biomass refers to any organic matter that has undergone the process of photosynthesis, wherein sunlight is absorbed and converted into chemical energy. The aforementioned sources encompass wood, energy crops, as well as agricultural as well as food waste. The concept of bioenergy encompasses the production of transportation fuels derived from organic substances. This section focuses on biomass and various biofuels, including biogas, ethanol, as well as biodiesel (Bioenergy, 2009).

### 3.3.1 Biomass

Biomass encompasses various forms of organic matter obtained from living and recently deceased organisms, which are utilized to generate heat, illumination, and culinary applications through combustion. The utilization of energy derived directly from biomass has been employed by human societies for numerous millennia and continues to account for approximately 8% of worldwide energy consumption. Bacteria have the ability to digest various forms of biomass, resulting in the production of biogas, predominantly composed of methane. This biogas can be harnessed for applications such as heating and power generation (Vassilev et al., 2010).

Various conversion processes can be employed to enhance the quality and practicality of biomass as a fuel. These processes encompass thermal, chemical, biochemical, as well as electrochemical mechanisms.

Thermal conversion processes primarily employ heat as the predominant mechanism for enhancing the quality of biomass. The fundamental options encompass torrefaction, pyrolysis, as well as gasification. Thermal gasification involves the application of heat to biomass while in the presence of oxygen, air, as well as water vapor. Pyrolysis refers to the process of thermally decomposing materials at high temperatures within an atmosphere devoid of reactive gases. This process entails a modification in the chemical structure and is characterized by its irreversibility. Torrefaction is a thermal process characterized by relatively low temperatures ranging from 200 to 300 °C, which represents a mild variant of pyrolysis (Goldemberg & Coelho, 2004). Chemical conversion processes, akin to those employed for fossil fuels like the Fischer-Tropsch synthesis, can be utilized to transform biomass into various commodity chemicals.

Biochemical conversion involves the utilization of microorganisms to facilitate the process of breaking down the molecular components of biomass. This encompasses various processes such as anaerobic digestion, fermentation, and composting.

The utilization of electrochemical oxidation presents a viable approach for converting directly of biomass into electrical energy. This can be executed using different categories of fuel cells, including direct carbon fuel cells, direct fuel from ethanol cells, direct formic acid fuel cells, L-ascorbic acid fuel cells, as well as microbial fuel cells (McKendry, 2002).

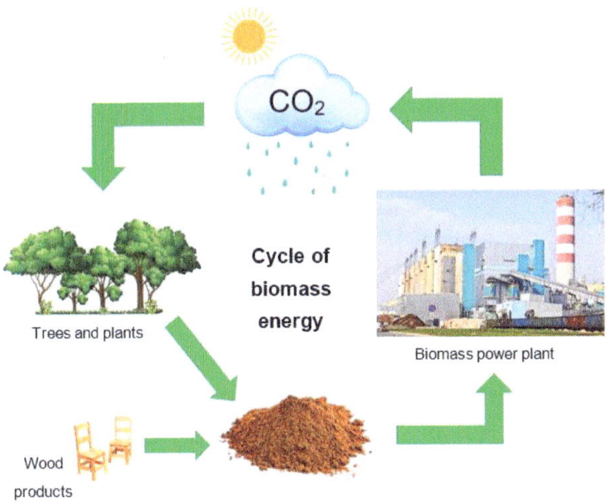


**Figure 3.2.** Illustration of factors that contribute to carbon-neutral bioenergy (Source: Kalak T. Potential Use of Industrial Biomass Waste as a Sustainable Energy Source in the Future. Energies. 2023; 16(4):1783. https://doi.org/10.3390/en16041783, Creative Commons License).

### 3.3.2 Biogas

Thermal gasification involves the application of heat to biomass while in the presence of oxygen, air, as well as water vapor. As a result, the molecular makeup of the material that is biological undergoes degradation, resulting in the generation of a gas mixture comprising different proportions of methane, carbon dioxide ($CO_2$), carbon monoxide (CO), hydrogen ($H_2$), and water vapor (Rasi et al., 2007).

Biological syngas is derived through the process of incomplete combustion, wherein the quantity of oxygen available is inadequate to facilitate complete combustion. This gas mixture consists of significant quantities of hydrogen ($H_2$) and carbon monoxide (CO) (Marchaim, 1992).

Pyrolysis, a process involving the application of heat in an oxygen-free or airless environment, results in the liberation of gas comprising $H_2$ light hydrocarbons such as methane or ethane, as well as liquid hydrocarbons.

Biogas is generated through the anaerobic digestion of organic matter by specific types of bacteria, such as mesophilic or thermophilic bacteria, in the absence of oxygen. This distinguishes biogas from biological syngas. The primary constituents of biogas consist of methane and carbon dioxide ($CO_2$).

There are two primary approaches utilized for the production of biogas: the extraction of biogas from landfills as well as the generation of biogas within enclosed vessels referred to as anaerobic digesters. Solid woody material is typically unsuitable for the production of biogas due to the presence of lignin, a compound that is resistant to anaerobic digestion by conventional bacterial species. Biogas exhibits promising prospects as an energy carrier, offering a diverse array of applications spanning from household utilization for cooking and heating purposes to large-scale electricity generation within the power grid (Deublein & Steinhauser, 2011).

### 3.3.3 Ethanol

Ethanol (C2H5OH) can be derived from basic sugars via the process of fermentation, which involves the biological conversion of sugars into alcohols, acids, and gases.

The chemical transformation for glucose is as follows (Galbe & Zacchi, 2002):

$$C_6H_{12}O_6 \rightarrow 2C_nH_5OH + 2CO_2 \qquad (3.2)$$

and for sucrose in the presence of water, we have

$$C_{12}H_{22}O_{11}\ \aleph H_2O \quad 4C_2H_5OH \quad 4CO_2 \qquad (3.3)$$

The density of energy of ethanol is roughly twice as much of simple sugars as a result of the liberation of carbon dioxide during the process of fermentation. Due to its liquid state under standard ambient conditions, ethanol possesses advantageous characteristics that make it a practical fuel for transportation and utilization purposes. The mass density of pure ethanol is recorded to be 0.789 kg/l, while its energy density is measured at 23.4 MJ/l. The energy density of ethanol, at 21.1 MJ/l, is significantly lower compared to that of gasoline (32 MJ/l). However, ethanol possesses a higher level of molecular stability and exhibits reduced knocking tendencies, as indicated by its octane number of 108.6 (Goldemberg, 2007).

The primary source of global ethanol production is derived from maize in the United States, with an annual output of 54 billion liters. Following closely is Brazil, which produces sugarcane-based ethanol at a volume of 24 billion liters per year. Both maize, as well as sugarcane, have the potential to serve as sources of food and animal feed. Consequently, scientific investigations are being conducted to explore the feasibility of producing ethanol from alternative organic materials. One example of these substances is cellulose, which serves as a fundamental component of plant structure. The development of a highly effective process for the production of cellulosic ethanol using cellulose materials derived from non-food plants would enable the utilization of a significant portion of the Earth's biomass for energy extraction purposes. Nevertheless, the production of cellulosic ethanol is currently in its nascent phase, with limited implementation restricted to small-scale experimental initiatives (Jones, 1989).

The Net Energy Balance (NEB) is a significant metric in assessing the viability of biofuels as an energy source, as it quantifies the ratio of output energy to input energy. For instance, utilizing contemporary agricultural techniques, the production of sugarcane ethanol may be achieved at an annual rate of 180 gigajoules per hectare per year. However, the production of ethanol necessitates significant energy inputs for activities such as fertilization, planting, and harvesting, as well as the subsequent transportation and distribution of ethanol. The findings of the analyses reveal that the net energy balance (NEB) for a typical Brazilian sugarcane plant is calculated as the ratio of total energy output to total energy input, resulting in a value of 8. The aforementioned scenario can be likened to the net energy balance estimates for maize-based ethanol derived from production plants in the United States, which fall within the range of 1.2 to

1.6. The inferior effectiveness of maize-based ethanol can be attributed to the higher energy requirements for processing the complex carbohydrates found in maize, specifically the starches amylopectin and amylose, which cannot be directly fermented. This stands in contrast to sugarcane-based ethanol, which exhibits more efficient processing. The determination of the net energy balance (NEB) for cellulosic production of ethanol remains uncertain; however, optimistic projections indicate that it could potentially surpass the NEB value observed for sugarcane ethanol (Lieber et al., 1989).

### 3.3.4 Biodiesel

Biodiesel is a liquid fuel derived from the chemical conversion of vegetable oil as well as animal fat. Fatty acid esters, derived from the processing of plant and animal oils, exhibit properties that closely resemble those of hydrocarbons found in traditional diesel fuels. The production of biodiesel from diverse feedstocks results in significant heterogeneity in the molecular composition across different batches of biodiesel. The energy density of biodiesel fuel is commonly observed to be approximately 37 MJ/kg, exhibiting a slight reduction of approximately 10% when compared to the energy density of conventional diesel fuel (Knothe, 2006).

The United States holds the top position, while Brazil occupies the second position, in terms of being the biggest exporters of fuel ethanol as well as biodiesel. In 2017, the aforementioned entities constituted 84% of the worldwide ethanol production and 26% of the biodiesel production. In both nations, the practice of blending fuel biodiesel and ethanol with conventional fossil transportation fuels is observed. Additionally, in Brazil, fuel ethanol is directly pitted against gasoline at fuel stations without any blending. Nevertheless, it is important to acknowledge that the production costs of biofuels, gasoline, as well as diesel exhibit variations across different countries. Consequently, these discrepancies in production costs have a direct impact on the breakeven oil price for both biodiesel and ethanol, thereby influencing the competitiveness of biofuels in both nations. Additionally, it is worth noting that policy measures also play a significant role in shaping the competitiveness of biofuels in these countries (Marchetti et al., 2007).

In order to evaluate the comparative competitiveness between biofuels and petroleum products, it is necessary to conduct a cost analysis comparing the production expenses of gasoline and diesel with those of ethanol as well as biodiesel. During the latter half of 2017, the production costs of gasoline as well as ethanol in Brazil amounted to 0.61 and 0.60 USD per liter, respectively. The respective expenses in the United States were 0.46 as well as 0.61 USD per liter. Comparable comparisons of prices for biodiesel and diesel in Brazil yield values of 0.58 and 0.80 USD/l, respectively, while in the United States, the corresponding figures are 0.47 and 0.83 USD/l (Demirbas, 2007).

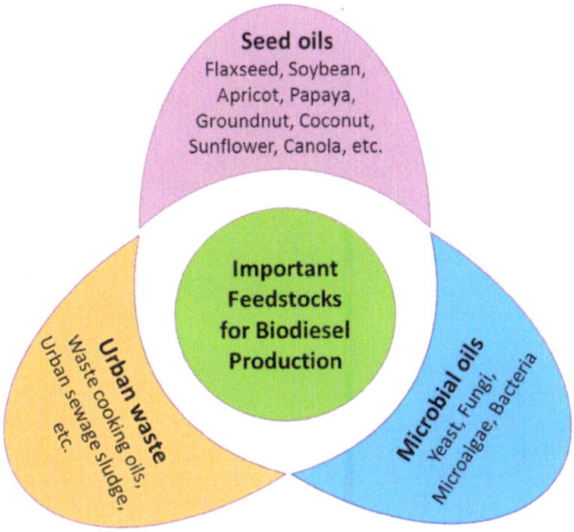


**Figure 3.3.**The lifecycle of the biodiesel (Source: Ingle AP, Chandel AK, Philippini R, Martiniano SE, da Silva SS. Advances in Nanocatalysts Mediated Biodiesel Production: A Critical Appraisal. Symmetry. 2020; 12(2):256. https://doi.org/10.3390/sym12020256, Creative Commons License).

## 3.4 FOSSIL FUELS

The primary fossil fuels which have propelled the advancement of contemporary society, and continue to hold significant prominence in human energy consumption, encompass coal, petroleum, as well as natural gas. This section provides a comprehensive overview of the fundamental attributes of fossil fuels (Welsby et al., 2021).

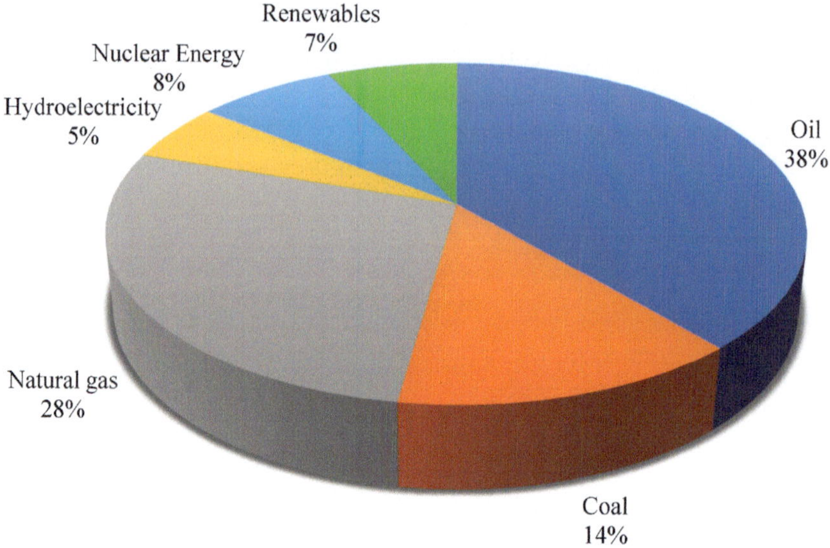


**Figure 3.4.** Representation of fossil fuel elements and their composition (Source: Hou H, Lu W, Liu B, Hassanein Z, Mahmood H, Khalid S. Exploring the Role of Fossil Fuels and Renewable Energy in Determining Environmental Sustainability: Evidence from OECD Countries. Sustainability. 2023; 15(3):2048. https://doi.org/10.3390/su15032048, Creative Commons License).

### 3.4.1 Coal

Coal is predominantly comprised of carbon (C), along with varying amounts of hydrogen (H), sulfur (S), oxygen (O), and nitrogen (O'Keefe et al., 2013).

### 3.4.2 Petroleum

The primary factor contributing to the prevalence of oil as a source of energy consumption stems from its advantageous characteristics, including its high energy density (42-44 MJ/kg) as well as its convenient transport and storage capabilities.

Crude oil is composed of a diverse array of hydrocarbons, including kinds of paraffin, cycloparaffins, olefins, as well as aromatic hydrocarbons, as referred to in the petroleum sector. The scientific nomenclature for ingredient molecules encompasses alkanes, cycloalkanes, alkenes, as well as arenes (Furimsky & Massoth, 2005).

Crude oil serves as a fundamental primary resource for the extraction of a range of products, including but not limited to naphtha, kerosene, gasoline, diesel, and heavier fuel oils, such as those utilized for residential heating purposes. In oil refineries, a method known as distillation is employed to segregate the different products. This involves subjecting crude oil to heat, causing the distinct chains to be extracted based on their respective vaporization temperatures (Magoon & Dow, 1994).

The hydrocarbon chains ranging from CH16 to C11H24 are combined and subsequently utilized in the formulation of gasoline. Subsequently, kerosene, falling within the carbon chain length of C12 to C15, is succeeded by diesel fuel as well as viscous fuel oils. Gasoline, also known as petrol, is a multifaceted blend comprising approximately 500 distinct molecules, which serves as a fuel source for motor vehicles. The fuel composition comprises hydrocarbons as well as non-hydrocarbon additives, which are carefully adjusted to facilitate elevated engine compression ratios and enhance engine efficiency, while simultaneously mitigating certain adverse environmental consequences (Bachmann et al., 2014).

### 3.4.3 Natural Gas

The majority of natural gas reserves consist predominantly of methane, accompanied by lesser proportions of ethane, propane, as well as butane, as well as trace amounts of higher hydrocarbons, nitrogen, carbon dioxide, water vapor, hydrogen, oxygen, sulfide, and occasionally helium (Li et al., 2016).

## 3.5 COMBUSTION

The primary method of harnessing energy from both biofuels as well as fossil fuels is predominantly achieved through the process of combustion. In more technical terms, combustion can be described as the chemical reaction that occurs when a substance undergoes rapid oxidation in the presence of oxygen, resulting in the release of heat and the formation of new chemical compounds. In broader terms, combustion can be described as an exothermic process in which a substance undergoes a complete reaction with an oxidizing agent. In the context of an ideal combustion process, it can be postulated that each carbon atom within the compound undergoes complete oxidation, resulting in the formation of carbon dioxide (CO2), while each hydrogen atom undergoes full oxidation, leading to the production of water (H2O). In practical applications, it is observed that combustion does not achieve complete oxidation. As an illustration, a portion of carbon may undergo complete combustion, resulting in the formation of ash, while another portion of carbon may undergo partial oxidation, leading to the production of carbon monoxide (CO) (Turns, 1996).

The stoichiometric combustion of a hydrocarbon in the presence of oxygen can be represented by the following chemical equation:

$$C_xH_y + \left(x + \frac{y}{4}\right)O_2 \rightarrow xCO_2 + \frac{y}{2}H_2O \tag{3.4}$$

Virtually all combustion processes occur by utilizing air as the primary source of oxygen. When considering all non-oxygen constituents in the air as nitrogen, the equation of combustion equation is modified as follows (Veynante & Vervisch, 2002):

$$C_xH_y + \left(x + \frac{y}{4}\right)O_2 + 3.77\left(x + \frac{y}{4}\right)N_2 \rightarrow xCO_2 + \frac{y}{2}H_2O + 3.77\left(x + \frac{y}{4}\right)N_2 \tag{3.5}$$

The nitrogen-to-oxygen volume ratio in this context is 3.77, as the volume fraction of oxygen in the air is 20.95%. Despite the fact that nitrogen does not actively partake in the combustion process, its presence in the flue gas provides valuable information regarding the overall composition.

Additional substances become noticeable at flame temperatures exceeding 1600 K, such as the potential oxidation of nitrogen to NO or $NO_2$, or incomplete combustion resulting in the presence of CO and $H_2$ molecules (Docquier & Candel, 2002).

### 3.5.1 Combustion of Gasoline

The comprehensive examination of gasoline combustion presents a challenging task owing to its intricate composition. To ensure ease of analysis, we make the assumption that the gasoline under consideration consists solely of isooctane (C8H18) with a purity of 100%. The reaction that represents the full combustion of two moles of isooctane with oxygen is as follows (Akal et al., 2020):

$$2C_8H_{18} + 25O_2 \rightarrow 18H_2O(l) + 16CO_2 \tag{3.6}$$

In this context, it is postulated that H2O exists in the liquid state after the combustion process. By applying Hess's law, the enthalpy of combustion for a single mole of isooctane can be determined.

$$\Delta I_c\left(C_8H_{18}\right) = \Delta I^f(CO_2(g)) + 9\Delta I^f(H_2O(l)) - \Delta I^f(C_8H_{18}) \tag{3.7}$$

By utilizing the given enthalpies of formation, namely $\Delta I^f$ $\Delta I^f$ ($CO_2CO_2$ (g)) =394 kJ/mol, ΔI f ($H_2H_2O$(l)) = 286 kJ/mol, and ΔI f (C8H18) =250 kJ/mol, , the resulting and $\Delta I_c \Delta I_c (C_8H_{18}C_8H_{18})$ =5.476 MJ/mol is determined to be 5.476 MJ/mol. Given the molar mass of iso-octane as 114 g/mol, it can be determined that its specific energy density, denoted as $E_c$, is equal to 47.5 MJ/kg, following the convention of positive values.

Gasoline, a blend of hydrocarbons, typically exhibits a heat of combustion (referred to as the negative of the greater heat value, HHV) of approximately 44 MJ/kg.

The ratio of fuel to air throughout the process of combustion holds significant importance. The mixture is considered stoichiometric when the amount of oxygen present is exactly sufficient to fully oxidize all hydrocarbons during combustion. The composition of the mixture can be characterized as lean when an excess of oxygen is present, while it is considered rich when there is an insufficient amount of oxygen to support complete combustion. The stoichiometric air-to-fuel ratio for gasoline is approximately 14.7:1 in terms of mass (Benajes et al., 2013).

### 3.5.2 Combustion of Ethanol

The chemical reaction that describes the combustion of ethanol in the presence of oxygen can be expressed as follows:

$$C_2H_5OH + 3O_2 \rightarrow 3H_2O(l) + 2CO_2 \quad (3.8)$$

In this context, it is postulated that H2O exists in a liquid state subsequent to the combustion process. By applying Hess's law, the enthalpy of combustion for a single mole of ethanol can be determined.

$$\Delta I_c \left(C_2H_5OH\right) = 2\Delta I^f (CO_2(g)) + 3\Delta I^f (H_2O(l)) - \Delta I^f (C_2H_5OH) \quad (3.9)$$

By utilizing the given enthalpies of formation, specifically, ΔI f ($CO_2$ (g)) = 394 kJ/mol, ΔI f ($H_2O$(l)) = 286 kJ/mol, and ΔI f ($C_2H_5OH$) = 278 kJ/mol, the resulting value for $\Delta I_c(C_2H_5OH)$ is determined to be 1.368 MJ/mol. The molar mass of ethanol, which is 46 g/mol, corresponds to a specific energy density (positive by convention) of $E_c$ = 29.7 MJ/kg. It is evident that ethanol possesses a smaller energy density in comparison to both isooctane as well as gasoline (Liu et al., 2021).

### 3.5.3 Combustion of Coal

Coal is combusted within power stations to generate electricity and produce valuable thermal energy. The coal is subjected to a crushing process in order to reduce its size to less than 5 cm, thereby facilitating its combustion. Subsequently, the coal is conveyed from the storage yard to the in-plant storage silos through a system of conveyor belts. In the context of pulverized coal combustion in plants, coal silos are responsible for supplying coal to pulverizers, also known as coal mills. These pulverizers are designed to take larger coal pieces, approximately 5 cm in size, and grind them down to a fine consistency resembling talcum powder. Subsequently, the pulverized coal is sorted and combined with primary combustion air. This air serves the dual purpose of transporting the coal to the boiler furnace as well as preheating it, thereby facilitating the removal of excessive moisture content from the coal. In a 500 MWe power plant, it is common to find a configuration consisting of six pulverizers. Among these pulverizers, five have the capacity to deliver coal to the furnace at a rate of 250 tonnes per hour when the plant is operating at full load.

The chemical reaction that characterizes coal combustion encompasses the elements carbon (C), sulfur (S), oxygen (O), nitrogen (N), hydrogen (H), as well as moisture ($H_2O$). Disregarding all non-combustible constituents, coal is denoted as C-H, and the combustion processes are approximated using the subsequent formulas.

$$2C - H + \frac{3}{2}O_2 \rightarrow H_2O + 2CO \text{ (incomplete)} \tag{3.10}$$

$$2C - H + \frac{5}{2}O_2 \rightarrow H_2O + 2CO_2 \text{ (complete)} \tag{3.11}$$

The energy content of coal typically falls within the range of Ec = 18–35 MJ/kg, although coal predominantly utilized for electricity generation exhibits an energy content of 20–25 MJ/kg (Wojtowicz et al., 1993).

#### *3.5.3.1 Fluidized Bed Combustion*

Since the 1980s, there has been ongoing development of Fluidized Bed Combustion (FBC) technology with the aim of mitigating plant emissions. This approach involves the upward passage of powerful air jets through coal, thereby maintaining the turbulent and fluid-like motion of the fuel particles during the process of combustion. The creation of nitrogen oxides ($NO_x$) is suppressed by combustion occurring at lower temperatures (typically 800-900 °C) due to enhanced heat transfer efficiency (Koornneef et al., 2007).

#### *3.5.3.2 Integrated Gasification Combined Cycle*

The Integrated Gasification Combined Cycle (IGCC) plant incorporates a coal gasification system within a combined cycle plant. The process of coal gasification commences with the utilization of steam reforming, whereby steam at elevated temperatures is introduced over a bed of coke, resulting in the formation of a combination of carbon dioxide and hydrogen monoxide, commonly referred to as water gas.

$$C + H_2O(g) \rightarrow H_2(g) + CO \tag{3.12}$$

The aforementioned reaction exhibits an endothermic nature, as indicated by a change in enthalpy (ΔIr) of 131 kJ/mol. In order for this reaction to occur, a temperature exceeding 1000 K is necessary. In the presence of excess steam, the water gas undergoes an exothermic reaction to yield hydrogen as well as carbon dioxide, as described by the water-gas shift reaction.

$$CO(g) + H_2O(g) \rightarrow H_2(g) + CO_2(g) \tag{3.13}$$

Consequently, the gas that is produced is a combination of hydrogen ($H_2$), carbon monoxide (CO), and carbon dioxide ($CO_2$). The entirety of the process is characterized by its complexity and high cost, yet it does yield certain enhancements in terms of cycle efficiency (Emun et al., 2012).

### 3.5.4 Combustion of Hydrogen

The chemical reaction involving the combination of hydrogen and oxygen results in the formation of water as the sole end product.

$$2H_2 + O_2 \rightarrow 2H_2O \tag{3.14}$$

This reaction may be achieved through either a direct process of combustion or in a fuel cell. The significance of the reaction (3.14) lies in its characteristic of emitting zero emissions, rendering it a highly sought-after alternative to the combustion of fossil fuels. Nevertheless, it should be noted that hydrogen exhibits lower energy density compared to alternative fuel sources. Furthermore, during the combustion process in the presence of atmospheric oxygen, the release of NOX particles is anticipated (Verhelst, 2005).

## SUMMARY

Biological as well as chemical energy are fundamental forms of energy that serve pivotal functions in diverse biological as well as physiological processes. Biological energy pertains to the energy that is stored and employed by living organisms in order to carry out vital physiological processes. Adenosine triphosphate (ATP) serves as the principal source of biological energy. Chemical energy refers to a form of potential energy that is contained within the chemical bonds of various compounds. Chemical reactions can result in the release or absorption of energy that is stored within the bonds. Chemical energy plays a crucial role in a multitude of natural phenomena and human endeavors. Fossil fuels, namely natural gas, coal, and oil, represent abundant reservoirs of chemical energy that have been derived from organic matter of ancient origin.

## REFERENCES

1. Akal, D., Öztuna, S., & Büyükakın, M. K. (2020). A review of hydrogen usage in internal combustion engines (gasoline-Lpg-diesel) from combustion performance aspect. International Journal of Hydrogen Energy, 45(60), 35257-35268.
2. Bach, L. T., Mackinder, L. C., Schulz, K. G., Wheeler, G., Schroeder, D. C., Brownlee, C., & Riebesell, U. (2013). Dissecting the impact of CO2 and pH on the mechanisms of photosynthesis and calcification in the coccolithophore Emiliania huxleyi. New Phytologist, 199(1), 121-134.
3. Bachmann, R. T., Johnson, A. C., & Edyvean, R. G. (2014). Biotechnology in the petroleum industry: an overview. International Biodeterioration & Biodegradation, 86, 225-237.
4. Benajes, J., García, A., Domenech, V., & Durrett, R. (2013). An investigation of partially premixed compression ignition combustion using gasoline and spark assistance. Applied Thermal Engineering, 52(2), 468-477.
5. Bioenergy, I. E. A. (2009). Bioenergy–a sustainable and reliable energy source. International Energy Agency Bioenergy, Paris, France.
6. Caplice, E., & Fitzgerald, G. F. (1999). Food fermentations: role of microorganisms in food production and preservation. International Journal of Food Microbiology, 50(1-2), 131-149.
7. Clayton, R. K. (1972). Physical mechanisms in photosynthesis: past elucidations and current problems. Proceedings of the National Academy of Sciences, 69(1), 44-49.
8. Demirbas, A. (2007). Importance of biodiesel as transportation fuel. Energy Policy, 35(9), 4661-4670.
9. Deublein, D., & Steinhauser, A. (2011). Biogas from waste and renewable resources: an introduction. John Wiley & Sons.
10. Docquier, N., & Candel, S. (2002). Combustion control and sensors: a review. Progress in energy and combustion science, 28(2), 107-150.
11. Eberhard, S., Finazzi, G., & Wollman, F. A. (2008). The dynamics of photosynthesis. Annual Review of Genetics, 42, 463-515.
12. Emerson, R., & Lewis, C. M. (1942). The photosynthetic efficiency of phycocyanin in Chroococcus, and the problem of carotenoid participation in photosynthesis. The Journal of General Physiology, 25(4), 579-595.
13. Emun, F., Gadalla, M., Majozi, T., & Boer, D. (2010). Integrated gasification combined cycle (IGCC) process simulation and optimization. Computers & Chemical Engineering, 34(3), 331-338.

14. Evans, J. R. (2013). Improving photosynthesis. Plant Physiology, 162(4), 1780-1793.
15. Farquhar, G. D., Von Caemmerer, S., & Berry, J. A. (2001). Models of photosynthesis. Plant Physiology, 125(1), 42-45.
16. Furimsky, E., & Massoth, F. E. (2005). Hydrodenitrogenation of petroleum. Catalysis Reviews, 47(3), 297-489.
17. Galbe, M., & Zacchi, G. (2002). A review of the production of ethanol from softwood. Applied Microbiology and Biotechnology, 59, 618-628.
18. Goldemberg, J. (2007). Ethanol for a sustainable energy future. Science, 315(5813), 808-810.
19. Goldemberg, J., & Coelho, S. T. (2004). Renewable energy—traditional biomass vs. modern biomass. Energy Policy, 32(6), 711-714.
20. Golueke, C. G., & Oswald, W. J. (1959). Biological conversion of light energy to the chemical energy of methane. Applied Microbiology, 7(4), 219-227.
21. González, M. C., Cejudo, F. J., Sahrawy, M., & Serrato, A. J. (2021). Current knowledge on mechanisms preventing photosynthesis redox imbalance in plants. Antioxidants, 10(11), 1789.
22. Harvey, M., & Pilgrim, S. (2011). The new competition for land: Food, energy, and climate change. Food Policy, 36, S40-S51.
23. Hu, G., Fan, Y., Zhang, L., Yuan, C., Wang, J., Li, W., ... & Li, F. (2013). Enhanced lipid productivity and photosynthesis efficiency in a Desmodesmus sp. mutant induced by heavy carbon ions. PloS One, 8(4), e60700.
24. Humphris, S. E., Zierenberg, R. A., Mullineaux, L. S., & Thomson, R. E. (1995). Seafloor hydrothermal systems: physical, chemical, biological, and geological interactions. Washington DC American Geophysical Union Geophysical Monograph Series, 91.
25. Jones, R. P. (1989). Biological principles for the effects of ethanol. Enzyme and Microbial Technology, 11(3), 130-153.
26. Kamerlin, S. C., & Warshel, A. (2010). The EVB as a quantitative tool for formulating simulations and analyzing biological and chemical reactions. Faraday Discussions, 145, 71-106.
27. Ke, B. (2001). Photosynthesis photobiochemistry and photobiophysics (Vol. 10). Springer Science & Business Media.
28. Knothe, G. (2006). Analyzing biodiesel: standards and other methods. Journal of the American Oil Chemists' Society, 83(10), 823-833.
29. Koornneef, J., Junginger, M., & Faaij, A. (2007). Development of fluidized

bed combustion—An overview of trends, performance and cost. Progress in energy and combustion science, 33(1), 19-55.

30. Li, X. S., Xu, C. G., Zhang, Y., Ruan, X. K., Li, G., & Wang, Y. (2016). Investigation into gas production from natural gas hydrate: A review. Applied Energy, 172, 286-322.
31. Lieber, C. S., Decarli, L. M., & Sorrell, M. F. (1989). Experimental methods of ethanol administration. Hepatology, 10(4), 501-510.
32. Liu, Y., Yue, L., Wang, Z., & Xing, B. (2019). Processes and mechanisms of photosynthesis augmented by engineered nanomaterials. Environmental Chemistry, 16(6), 430-445.
33. Liu, Z., Sun, P., Du, Y., Yu, X., Dong, W., & Zhou, J. (2021). Improvement of combustion and emission by combined combustion of ethanol premix and gasoline direct injection in SI engine. Fuel, 292, 120403.
34. Magoon, L. B., & Dow, W. G. (1994). The petroleum system: Chapter 1: Part I. Introduction.
35. Marchaim, U. (1992). Biogas processes for sustainable development (No. 95-96). Food & Agriculture Org..
36. Marchetti, J. M., Miguel, V. U., & Errazu, A. F. (2007). Possible methods for biodiesel production. Renewable and Sustainable Energy Reviews, 11(6), 1300-1311.
37. McKendry, P. (2002). Energy production from biomass (part 1): Overview of biomass. Bioresource Technology, 83(1), 37-46.
38. O'Keefe, J. M., Bechtel, A., Christanis, K., Dai, S., DiMichele, W. A., Eble, C. F., ... & Hower, J. C. (2013). On the fundamental difference between coal rank and coal type. International Journal of Coal Geology, 118, 58-87.
39. Qian, H. (2007). Phosphorylation energy hypothesis: open chemical systems and their biological functions. Annu. Rev. Phys. Chem., 58, 113-142.
40. Rasi, S., Veijanen, A., & Rintala, J. (2007). Trace compounds of biogas from different biogas production plants. Energy, 32(8), 1375-1380.
41. Raven, J. A., Cockell, C. S., & De La Rocha, C. L. (2008). The evolution of inorganic carbon concentrating mechanisms in photosynthesis. Philosophical Transactions of the Royal Society B: Biological Sciences, 363(1504), 2641-2650.
42. Simkin, A. J., López-Calcagno, P. E., & Raines, C. A. (2019). Feeding the world: improving photosynthetic efficiency for sustainable crop production. Journal of Experimental Botany, 70(4), 1119-1140.

43. Souri, M. K. (2016). Aminochelate fertilizers: the new approach to the old problem; a review. Open Agriculture, 1(1), 118-123.
44. Turns, S. R. (1996). Introduction to combustion (Vol. 287, p. 569). New York, NY, USA: McGraw-Hill Companies.
45. Vassilev, S. V., Baxter, D., Andersen, L. K., & Vassileva, C. G. (2010). An overview of the chemical composition of biomass. Fuel, 89(5), 913-933.
46. Verhelst, S. (2005). A study of the combustion in hydrogen-fuelled internal combustion engines.
47. Veynante, D., & Vervisch, L. (2002). Turbulent combustion modeling. Progress in energy and combustion science, 28(3), 193-266.
48. Welsby, D., Price, J., Pye, S., & Ekins, P. (2021). Unextractable fossil fuels in a 1.5° C world. Nature, 597(7875), 230-234.
49. Wojtowicz, M. A., Pels, J. R., & Moulijn, J. A. (1993). Combustion of coal as a source of N2O emission. Fuel Processing Technology, 34(1), 1-71.

CHAPTER

# 4 Mechanical and Electromagnetic Energy

## Contents

# INTRODUCTION

Mechanical and electromagnetic energy are two prominent forms of energy that significantly impact various aspects of our daily existence. The transportation of individuals, as well as the conveyance of food, raw materials, and various commodities, necessitates the utilization of mechanical energy. In 2016, approximately 31.6% of the global Total Final Consumption (TFC) of energy was allocated towards transportation, encompassing personal and commercial means of land, water, and air transport. In transportation, all vehicles must be able to store and carry their fuel unless they can connect to an external energy-supplying grid. This requirement exhibits a preference for using fossil fuels, specifically gasoline, due to its notable energy density and ease of combustion. Hence, it can be observed that a significant proportion of the ultimate utilization of oil products, specifically 65.1%, is allocated towards transportation purposes (Lai et al., 2021).

Similarly, a notable fraction of the ultimate utilization of natural gas, amounting to 7.1%, is also dedicated to transportation. Therefore, it is imperative to investigate the mechanisms by which energy is utilized to initiate and sustain the movement of a vehicle (such as a car on the road or an airplane in flight), with the potential goal of minimizing energy requirements regardless of the chosen fuel source. This chapter will examine the processes involved in the generation and dissipation of mechanical energy, as well as the mechanisms by which mechanical energy can be stored (Jia et al., 2017).

Electromagnetic energy has become progressively significant in our everyday energy consumption. In 2016, global electricity consumption amounted to 20,863 terawatt-hours (or 5.795 exajoules), with industrial usage accounting for 41.6%, residential usage accounting for 27.2%, and commercial usage accounting for 22%. The electricity utilization in the agricultural sector accounted for 2.9%, while the transportation sector accounted for 1.7%. The utilization of electromagnetic energy-based devices is prevalent and anticipated to expand further due to their cleanliness and convenience. This chapter explores the fundamental aspects of electromagnetic energy, which play a crucial role in its generation, transmission, storage, and utilization (Mei et al., 2019).

## 4.1 FORCES AND FIELDS

The classical law of dynamics, formulated by Sir Isaac Newton, establishes a fundamental connection between the force exerted on an object, denoted as F, and its resulting acceleration, represented by $a = dv/dt$, where v is the object's velocity and t is time. This relationship applies to objects with mass m. The second law of dynamics, as formulated by Sir Isaac Newton, mathematically expresses the relationship between force (F), mass (m), and acceleration (a) as F = ma. It is currently understood that this relationship holds solely under the condition of a constant mass of the object. Nevertheless, as demonstrated by Einstein, the mass m is subject to variation in accordance with its velocity. However, it is important to note that the impact of this effect is exceedingly minute in our daily lives, and the Newtonian formulation can be utilized for most of the scenarios examined within this text (Davis, 1964).

Therefore, the influence of a force exerted on an object of a specific mass is intricately connected to alterations in its velocity and, consequently, to variations in its kinetic energy. Among the various forces in our surrounding world, it can be argued that gravitation and electromagnetism hold the utmost significance. These forces exhibit a close correlation with the gravitational and electromagnetic fields. This section provides a brief overview of the fields and their corresponding associated forces (Berger, 1985).

### 4.1.1 A Force

A force can be defined as the exertion of influence by one object or field upon another object. The second law of motion, as formulated by Newton, delineates the relationship between the force exerted on an object, denoted as F, and the object's mass, represented by m (Gutowski & Force, 2009).

$$\mathbf{F} = \frac{\mathrm{d}\mathbf{p}}{\mathrm{d}t} = \frac{\mathrm{d}(m\mathbf{v})}{\mathrm{d}t}. \qquad (4.1)$$

The equation, initially formulated by Newton, posits that the derivative of an object's linear momentum p with respect to time is equal to the force F exerted on the object. Equation (4.1) is universally applicable and represents a fundamental principle inherent in the laws of nature. The validity of this statement was verified by Einstein's theory of special relativity, which demonstrates that the mass of an object is not fixed but rather fluctuates in proportion to its velocity, denoted as v. Nevertheless, in the case of objects traveling at velocities significantly lower than the speed of light, the alteration in mass is exceedingly minute and can be disregarded (Postlethwait et al., 1998). In instances of one-dimensional motion, the second law of dynamics is expressed as follows,

$$F = m\frac{d^2x(t)}{dt^2} = m\ddot{x}, \tag{4.2}$$

The shorthand notation $x \equiv d^2x(t)/dt^2$ is commonly employed in academic discourse to represent the concept of acceleration. The variable x(t) denotes the position of an object with mass m at a given time t.

The two primary forces that arise from the distant interactions of gravitational and electromagnetic nature between celestial bodies can be described as follows,

$$\mathbf{F}_g = -\frac{G_N m_1 m_2}{r^2}\hat{\mathbf{r}}, \tag{4.3}$$

The gravitational attraction force, denoted as $F_g$, is exerted by mass $m_1$ on mass $m_2$. This force is directed from mass $m_2$ towards mass $m_1$. The distance between the centers of the masses is represented by r. The unit vector r-hat indicates the direction from mass $m_1$ to mass $m_2$. The value of the gravitational constant, denoted as $G_N$, is approximately $6.67430(15) \times 10^{11}$ $m^3kg^{-1}$. The symbol s^2 represents the Newtonian constant of gravitation (Taddeo & Floridi, 2018).

$$\mathbf{F}_{em} = \frac{1}{4\pi\varepsilon_0}\frac{q_1 q_2}{r^2}\hat{\mathbf{r}}, \tag{4.4}$$

The variable "Fem" represents the electromagnetic force generated by a charged body, denoted as $q_1$, on another charged body, denoted as $q_2$. The unit vector "r" indicates the direction from charge $q_1$ to charge $q_2$, while $\varepsilon_o$ represents the electric permittivity of a vacuum.

Based on empirical observations, it has been established that the presence of charges in motion gives rise to an additional force. The force in question can be conveniently characterized by its association with the magnetic field (Amores et al., 1998).

### 4.1.2 A Field

A field refers to a localized mechanism that facilitates the interaction of forces. The gravitational field is accountable for exerting the gravitational force (4.3), while the electric field is the medium for force transmission (4.4). Magnetic fields are observed when charges are in motion, resulting in the manifestation of the magnetic force (Vanommeslaeghe et al., 2010). The electric field is characterized by a vector E(x, t) at each spatial point x and temporal moment t. Firstly, we examine a static electric field denoted as E(x). This particular domain applies a force denoted as F on a charge q, as provided,

$$\mathbf{F} = q\mathbf{E}. \tag{4.5}$$

Given that the electric field arises due to a charge Q at point $x_o$, we compare Equations (4.4) and (4.5).

$$\mathbf{E} = \frac{Q}{4\pi\varepsilon_0 r^2}\hat{\mathbf{r}}, \tag{4.6}$$

The expression for the electric field at a specific location x, resulting from a collection of charges $Q_i$ situated at points $x_i$, can be obtained by superposing the individual electric fields. In this context, r represents the distance between the location x and a reference point $X_{o,}$ given by the absolute value of the difference between X and $X_o$.

$$\mathbf{E} = \frac{1}{4\pi\varepsilon_0}\sum_i Q_i \frac{\mathbf{x} - \mathbf{x}_i}{|\mathbf{x} - \mathbf{x}_i|^3}. \tag{4.7}$$

Graphical representations of the electric field are commonly depicted using field lines that exhibit parallelism with the direction of the electric field vector E. As depicted in Figure 4.1, the field lines originate from positively charged entities and terminate at negatively charged entities (Liwo et al., 2019).

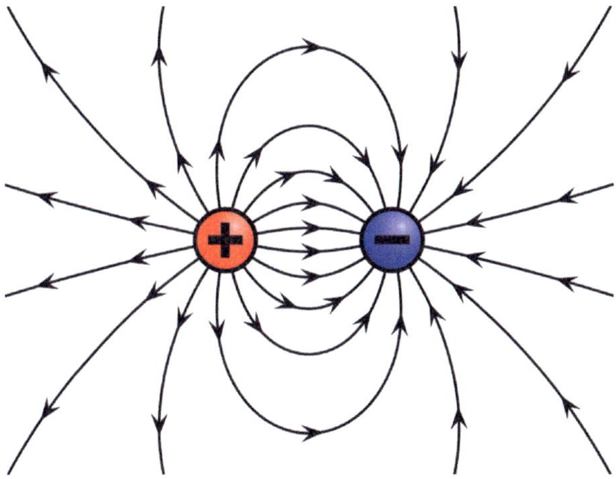

**Figure 4.1.** Representation of electric field produced by a pair of two opposite charges (left) and a pair of two negative charges (right). (Source: By Geek3 - Own work, CC BY-SA 3.0, https://commons.wikimedia.org/w/index.php?curid=10568742, Creative Commons License).

## 4.2 MECHANICAL ENERGY

The mechanical energy of an object arises from two distinct sources: its motion, known as kinetic energy, and its position within fields, referred to as potential energy. This form of energy holds particular significance in the context of transportation, as it is necessary for initiating the movement of a vehicle. This section will examine the two fundamental forms of mechanical energy: kinetic and potential (Maamer et al., 2019).

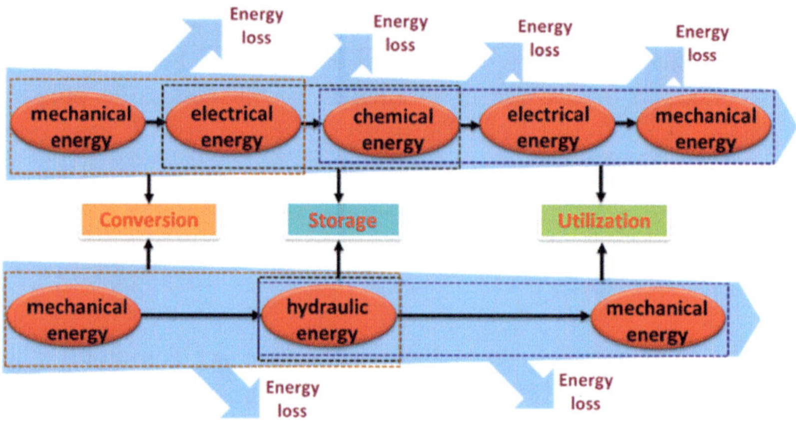


**Figure 4.2.** Illustration of the definition of mechanical energy with example (Source: Shi H, Liu Z, Mei X. Overview of Human Walking Induced Energy Harvesting Technologies and Its Possibility for Walking Robotics. Energies. 2020; 13(1):86. https://doi.org/10.3390/en13010086, Creative Commons License).

### 4.2.1 Kinetic Energy

The kinetic energy of an object of mass m moving with speed v is given by

$$E_K = \frac{1}{2}mv^2. \tag{4.8}$$

### 4.2.2 Potential Energy

Potential energy refers to the energy that is stored within a system of objects that engage in interactions governed by forces. The equation that describes the change in potential energy denoted as $dE_p$, for a particle undergoing a displacement dx in a field influenced by a force F, is expressed as follows (Schlegel, 2003):

$$dE_P = -dE_K = -\mathbf{F} \cdot \mathbf{dx}, \tag{4.9}$$

Where Ek represents the respective alteration in kinetic energy. The symbols within the equation imply that the potential energy exhibits an increase when the particle undergoes motion in a direction opposite to the direction of the force. If the condition above holds true, the integral of F dx is less than zero, implying that the partial derivative of E concerning p is greater than zero. The gravitational force acting on a body with mass m in the vicinity of Earth's surface within Earth's gravitational field can be represented by the equation F = mg. The body transitions from the Earth's surface to a position z units above it; its potential energy experiences an increase (Lorenz, 1955).

$$E_P(z) = \int_0^z dE_P = \int_0^z mg dz' = mgz. \tag{4.10}$$

It can be observed that the initial potential energy, denoted as $E_p(0)$, is equal to zero. Furthermore, the gravitational potential energy exhibits a linear relationship with respect to the height. The integral expressed in Equation (4.10) exhibits path independence. Forces for which the integral (4.10) exhibits path independence are referred to as conservative forces, as they uphold the conservation of the combined kinetic and potential energy. Non-conservative forces, commonly called dissipative forces, encompass friction and the electromotive force generated by a varying magnetic field (Braams & Bowman, 2009).

Using equations (4.10) and (4.3), the gravitational potential energy of a body with mass m, situated at a center-to-center distance r = |r| from a body with mass M, can be determined.

$$E_P(r) = -\frac{G_N M m}{r}. \tag{4.11}$$

The reference point for the potential energy in this context is zero when the distance r approaches infinity. As the objects converge, their kinetic energy increases while their potential energy experiences a corresponding decrement (Sato, 1955).

**Figure 4.3.** Representation of potential, kinetic, and mechanical energy (Source: Daniel Cole, Creative Commons License).

### 4.2.3 Work and Power

The term "work" refers to the energy that is transferred to an object as a result of the application of a force. The energy of an object increases when an external force performs positive work on it. Energy is dissipated from an object when negative work is exerted upon it or when the object performs work on its surrounding environment. When an object undergoes displacement represented by the differential distance vector dx while being subjected to the influence of a force F, the differential work dL performed by force can be described as follows (Tew, 2006):

$$\mathrm{d}L = \mathbf{F} \cdot d\mathbf{x} = F\mathrm{d}x\cos(\mathbf{F}, \mathbf{dx}), \tag{4.12}$$

The expression "cos (F, dx)" represents the cosine of the angle formed by the vectors F and dx. It should be noted that the work done is equal to zero in cases where the vectors F and dx are orthogonal to each other.

The term "power" describes the rate at which work is performed. In our possession, we possess

$$N = \frac{\mathrm{d}L}{\mathrm{d}t} = \mathbf{F} \cdot \frac{\mathbf{dx}}{\mathrm{d}t} = Fv\cos(\mathbf{F}, \mathbf{v}), \tag{4.13}$$

where N is the power, v = dx/dt is the velocity vector and v = |v| (Bell & Taylor, 2003).

### 4.2.4 Linear and Angular Momentum

The linear momentum of an object of mass m moving with velocity v is defined as

$$\mathbf{p} \equiv m\mathbf{v}. \tag{4.14}$$

For a system containing several constituents with certain momenta pi, the total momentum of the system is is $p_{tot} = \sum p_i$. For an isolated system the total momentum $p_{tot}$ $p_{tot}$ is conserved. The kinetic energy of an object with mass m and momentum p is $E_k = |p|^2 / (2m)$ (Kajita et al., 2003).

The angular momentum of a solitary particle, in relation to an arbitrary immobile point (within an inertial frame of reference) and possessing a linear momentum p, is formally characterized as

$$\mathbf{L} \equiv \mathbf{r} \times \mathbf{p}, \tag{4.15}$$

The variable "r" represents the vector that denotes the distance between a fixed point and the particle's location. The kilogram-meter squared per second ($kgm^2/s$) is recognized as the standard unit of measurement for angular momentum in the International System of Units (SI) (Belinfante, 1940).

A torque T is also defined with respect to the same point.

$$\mathbf{T} \equiv \mathbf{r} \times \mathbf{F}, \tag{4.16}$$

The symbol F represents the force exerted on the particle. The SI unit for torque is expressed as Newton meter (N m). The time differential of the angular momentum, as expressed in Equation (4.16), can be described as follows.

$$\frac{d\mathbf{L}}{dt} = \frac{d\mathbf{r}}{dt} \times \mathbf{p} + \mathbf{r} \times \frac{d\mathbf{p}}{dt} = \mathbf{r} \times \frac{d\mathbf{p}}{dt} = \mathbf{r} \times \mathbf{F} = \mathbf{T}, \tag{4.17}$$

Since

$$\frac{d\mathbf{r}}{dt} \times \mathbf{p} = \mathbf{v} \times m\mathbf{v} = 0. \tag{4.18}$$

Thus we obtained an important result

$$\frac{d\mathbf{L}}{dt} = \mathbf{T}. \quad (4.19)$$

The equation presented above asserts that the derivative of the angular momentum of an object is equivalent to the torque exerted on said object. The equation presented is synonymous with the second law of dynamics, as denoted by Equation (4.1), and is commonly referred to as the principle of angular momentum conservation. It is important to acknowledge that this conservation law holds for particles that follow closed orbits and has broader applicability (Miller & Munro, 1985).

When a particle located at a distance r from the origin undergoes rotational motion with a constant angular velocity ω around the origin, its linear speed can be expressed as $v = r\omega$, and its centripetal acceleration can be represented as $a = -\omega^2 r$. The linear momentum of a revolving particle can be expressed as $p = mv = mr\omega$, where m represents the particle's mass, v represents its linear velocity, r represents the radius of its circular path, and ω represents its angular velocity. Similarly, the angular momentum can be calculated as $L = rp = mr^2\ \omega$, where $r^2$ represents the radius square (Harp & Berne, 1968).

### 4.2.5 Mechanical Energy Losses

In conjunction with the aforementioned conservative forces, dissipative forces facilitate the conversion of kinetic energy into thermal energy, characterized by the stochastic movement of molecules.

- This category encompasses the friction force that arises between two bodies in direct contact and moving at different velocities.
- The drag force occurs on a solid surface when it is in motion relative to a fluid, such as when a car moves through the air.
- Rolling resistance occurs when a solid object, such as a car tire, rolls on the surface of another object, such as a road.

In a rigid body, such as a car, or a non-rigid body, such as a liquid droplet, in motion within a surrounding fluid, a force is generated in the opposite direction of the relative velocity vector between the body and the fluid. The force in question is commonly referred to as the drag force, which the following equation can mathematically represent (Heinrich et al., 1996):

$$\mathbf{F}_D = -\frac{1}{2} C_D A \rho v^2 \hat{\mathbf{v}}, \quad (4.20)$$

The equation provided involves several variables. The variable v represents the unit relative velocity vector, A represents the reference area of the body (typically defined as the cross-sectional area of the body in a plane perpendicular to the velocity vector v), ρ represents the density of the fluid, and CD represents the drag coefficient. The drag coefficient is contingent upon the geometric configuration of the object as well as the magnitude of its velocity relative to the surrounding fluid. The determination of this coefficient is typically conducted through experimental means, with its value obtained from Equation (4.21) after measuring all other parameters involved in the equation. As an illustration, contemporary passenger cars typically exhibit a drag coefficient ranging from 0.25 to 0.30, while conventional passenger aircraft tend to possess a drag coefficient of approximately 0.03 (Biewener, 2006).

Upon determining the drag force, the energy imparted to the fluid by the displacement of the body by a differential distance dx can be expressed as dE = FD dx. Consequently, the rate at which the body dissipates its kinetic energy can be represented as dE/dt = FD dx/dt = FD v. The process of evaluating this mathematical expression by performing integration over the distance D that the body has traveled within a given time period T results in the following outcome.

$$\Delta E = \int_0^T (\mathrm{d}E/\mathrm{d}t)\mathrm{d}t = \int_0^D \mathbf{F}_D \cdot \mathrm{d}\mathbf{x} = \frac{1}{2}C_D A D \rho v^2. \qquad (4.21)$$

It is observed that the energy dissipation resulting from the drag force varies proportionally to the square of the relative velocity. This paper aims to provide a comprehensive analysis of the role of fluid mechanics in energy transformation across various domains (Venkatachari et al., 2007).

### 4.2.6 Mechanical Energy Storage

Efforts have been undertaken to devise techniques for the storage of mechanical energy. The predominant approach entails utilizing pumped-storage hydropower (PSH), which involves the utilization of two water reservoirs situated at varying elevations. Pumped storage hydroelectricity (PSH) can expend energy by transferring water from a lower reservoir to an upper reservoir. The retrieval of potential energy from the upper reservoir occurs as water flows downward through a turbine. The overall efficiency of the pumped storage hydroelectricity (PSH) system ranges from 65% to 70%.

The pumped storage method necessitates expansive land areas and a particular topography. A flywheel storage system may be deemed suitable for applications that require a more compact solution. Flywheels are mechanical devices specifically engineered to store and retain rotational kinetic energy effectively. Objects resist changes in rotational speed due to their moment of inertia. The expression for the stored rotational kinetic energy is provided as follows:

$$E_{rot} = \frac{1}{2} I\omega^2, \quad (4.22)$$

Where, I is the moment of inertia of a rotating body (expressed in kg $m^2$ $m^2$ ) and ω is the angular velocity (expressed in rad/s) (Mahmoud et al., 2020).

## 4.3 ELECTROMAGNETIC ENERGY

Electromagnetic energy arises due to the interactions between electrically charged objects, which give rise to forces and fields. The utilization of electromagnetic energy by humans can be categorized into four distinct stages: generation, transmission, storage, and utilization.

Compact, clean, and convenient are notable characteristics of devices that utilize electromagnetic energy. The transmission of electromagnetic energy exhibits the ability to traverse long distances with minimal energy dissipation efficiently.

The conversion of mechanical energy to electrical energy, and vice versa, can be achieved with high efficiency by utilizing electric motors and generators. On the other hand, the conversion of thermal or chemical energy into electromagnetic energy is associated with substantial energy losses.

### 4.3.1 Electrostatics

Electrostatics is a subfield of physics that investigates the behavior and properties of stationary electric charges. Electrostatics is a branch of physics that focuses on studying electric fields, electric forces, and electric potentials that arise due to the presence of stationary electric charges.

The electric field, which arises from a stationary charge ρ(x) with units of $C/m^3$, and is uniformly spread across all regions of space, obeys the subsequent equations (Walker et al., 2011):

$$E_{rot} = \frac{1}{2} I\omega^2, \tag{4.23}$$

And,

$$\nabla \times \mathbf{E}(\mathbf{x}) = 0. \tag{4.24}$$

Both of these equations are adequate for determining the electric field E(x) that arises from any static charge distribution in a unique manner. Once the electric field is determined, the electrostatic voltage difference between two points $x_1$ and $x_{2,}$ can be calculated.

$$V(\mathbf{x}_2) - V(\mathbf{x}_1) = -\int_{\mathbf{x}_1}^{\mathbf{x}_2} \mathbf{E} \cdot d\mathbf{x}. \tag{4.25}$$

The magnitude of the force experienced by a charge q due to an electric field can be determined using the equation (4.5). The potential denoted by Equation (4.28) is quantified in terms of energy per unit charge, known as volts. It is worth noting that 1 volt is equivalent to 1 joule per coulomb (Honig & Nicholls, 1995).

A conductor refers to a substance that possesses charges capable of unrestricted mobility. Metals, including iron, aluminum, and copper, possess the characteristic of being conductive. On the other hand, materials that exhibit the property of inhibiting the free movement of charges, such as rubber, glass, and wood, are commonly referred to as insulators.

A capacitor is an electrical component that consists of two distinct conductors. The process of charging a capacitor involves establishing a potential difference by connecting its two conducting elements to a battery. The expression for the charge on the capacitor can be determined as

$$Q = CV, \tag{4.26}$$

The symbol C represents the quantity known as capacitance, measured in the SI unit farad (F). Notably, 1 farad (F) equals 1 coulomb per volt (C/V). Equation (4.26) asserts a direct proportionality between the charge on the capacitor and the voltage. The capacitance of a classic parallel plate capacitor can be expressed as the ratio of the electric charge stored on one of the conducting plates to the potential difference between the plates. The area of the plates determines it, the distance between them, and the dielectric material present in the gap,

$$C = \kappa \varepsilon_0 A / d, \tag{4.27}$$

Where $\varepsilon_0$ $\square_0$ is the permittivity of the vacuum, and κ is the material-dependent dielectric constant (Cevc, 1990).

### 4.3.2 Electric Current

Electric currents in electric circuits are quantified as the overall rate at which charges traverse a specific location within a wire.

$$I = \frac{dQ}{dt}. \tag{4.28}$$

The electric current is quantified in amperes within the International System of Units (SI), representing one of its fundamental units.

The current flow (I) through a conductor is directly proportional to the voltage drop (V) across the conductor, where the resistance (R) of the conductor serves as the constant of proportionality.

$$V = IR. \tag{4.29}$$

The association above is commonly referred to as Ohm's law. The unit of electrical resistance in the International System of Units (SI) is the ohm, denoted by the symbol Ω. It is defined as the voltage (V) to current (A) ratio, such that 1 ohm equals 1 volt per ampere. Typically, the temporal dependence of the voltage and current in Equation (4.26) can be observed.

When an electric charge is passed through a resistor, the energy expended is dissipated as a result of the interactions between the electrons and the constituent molecules of the resistor's material. Consequently, the conversion of electromagnetic energy into thermal energy within a resistor is commonly referred to as Joule heating, which denotes the dissipation of power in said resistor. The determination of the power dissipation rate, denoted as N Joule(t), is governed by Joule's law in the following manner.

$$N_{Joule}(t) = V(t)I(t) = I^2(t)R = \frac{V^2(t)}{R}, \tag{4.30}$$

The application of Ohm's law in various contexts. As elucidated by Joule's law, resistive heating exhibits a near-perfect efficiency of approximately 100% and is commonly employed in various domestic appliances. Nevertheless, in numerous applications, Joule heating is a significant contributor to energy dissipation (Munir et al., 2011).

### 4.3.3 Magnetism

Magnetic phenomena are integral to converting electrical energy into mechanical energy through the utilization of motors, as well as the subsequent reconversion of mechanical energy back into electrical energy via generators. The motion of electrically charged particles generates a magnetic field, which in turn applies a force on other moving charged particles. Like the electric field, the magnetic field B(x, t) can be characterized as a vector quantity defined at a specific point in space and time. The force exerted on an electric charge, denoted as q, that is in motion with a velocity v within a magnetic field can be described as follows (Stöhr & Siegmann, 2006):

$$\mathbf{F} = q\mathbf{v} \times \mathbf{B}. \tag{4.31}$$

The Lorentz force law describes the combined effects of stationary, space-dependent electric and magnetic fields on a charge q,

$$\mathbf{F}(\mathbf{x}) = q\mathbf{E}(\mathbf{x}) + q\mathbf{v} \times \mathbf{B}(\mathbf{x}). \tag{4.32}$$

Due to the absence of elementary magnetic charges, the magnetic field's equivalent to Eq. (4.29) can be expressed as follows.

$$\nabla \cdot \mathbf{B}(\mathbf{x}) = 0. \tag{4.33}$$

A notable implication arising from the structure of Equation (4.50) is that the lines of the magnetic field lack any starting or ending points, necessitating their configuration as closed curves.

The validity of Ampere's law in the given form holds true for a stationary current and charge density.

$$\nabla \times \mathbf{B}(\mathbf{x}) = \mu_0 \mathbf{j}(\mathbf{x}), \tag{4.34}$$

The current density, denoted as j(x), is determined by the local formulation of Ohm's law. The constant $\mu_0$, which is present in Equation (4.34), is referred to as the magnetic permeability of the vacuum.

Equations (4.33) and (4.34) provide a unique determination of the magnetic field that is linked to any current distribution that remains constant over time. The magnetic field surrounding a lengthy, linear conductor with a consistent electric current I can be determined as follows.

$$\mathbf{B}_{wire} = \frac{\mu_0}{2\pi} I \frac{\hat{\mathbf{z}} \times \hat{\rho}}{\rho}, \tag{4.35}$$

The variable $\hat{z}$ $\hat{z}$ represents a unit vector that indicates the direction along the wire, while the variable $\hat{\rho}$ $\hat{\rho}$ represents a vector that points radially outward from the wire. Applying the right-hand rule for the cross product between the unit vector in the z-direction and the unit vector in the radial direction results in the fingers of the right-hand wrapping in the direction of the magnetic field. This occurs when the thumb aligns with the direction of the current flow within the wire.

An additional illustration of a basic magnetic field can be observed in the solenoid, a structure consisting of a lengthy wire coiled in a spiral manner around a cylinder with a cross-sectional area denoted as A and a length denoted as l. The cylinder has the potential to either contain no substance or be occupied by a material with a permeability denoted as μ, such as iron, where the value of μ is significantly greater than the permeability of free space, $\mu_0$. The magnetic field within a solenoid is oriented parallel to the axis of the solenoid, and the following equation can determine its magnitude:

$$\mathbf{B}_{sol} = \mu n I \hat{\mathbf{z}}, \qquad (4.36)$$

Let n be defined as the ratio of the number of turns of wire N to the length l along the solenoid. Let I represent the current flowing through the solenoid, and let $\hat{z}$ denote a unit vector aligned with the solenoid axis, pointing in the direction of positive current (Verosub & Roberts, 1995).

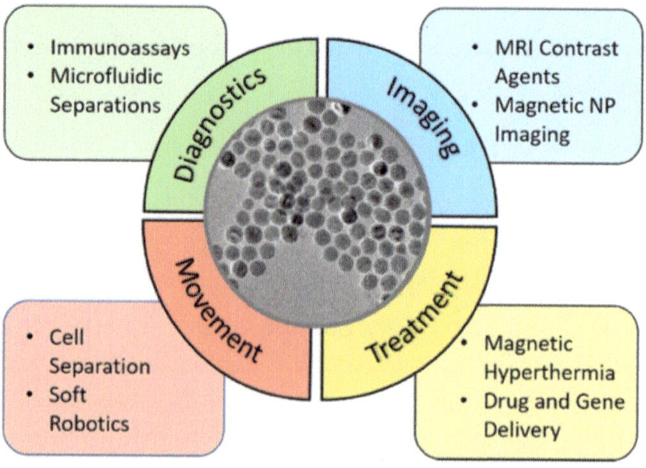


**Figure 4.4.** Illustration of the utilization of magnetism in various applications (Source: Stueber DD, Villanova J, Aponte I, Xiao Z, Colvin

VL. Magnetic Nanoparticles in Biology and Medicine: Past, Present, and Future Trends. Pharmaceutics. 2021; 13(7):943. https://doi.org/10.3390/pharmaceutics13070943, Creative Commons License).

### 4.3.4 Induction

Electric fields are generated as a result of alterations in magnetic fields, and conversely, magnetic fields are produced when electric fields undergo changes. There are two distinct methods for generating an electric field, which are outlined below:

- When a wire loop is subjected to a magnetic field that is increasing in magnitude.
- When a wire loop is affixed within a spatially constant magnetic field that exhibits temporal growth.

Faraday's law of induction determines the magnitude of the generated electric field, a principle closely associated with the underlying mechanisms of electric motors and generators, which will be further explored in the subsequent subsection. The differential form of Faraday's law of induction is expressed as follows (Cohen, 1995).

$$\nabla \times \mathbf{E} = -\frac{\partial \mathbf{B}}{\partial t}, \qquad (4.37)$$

Or in the integral form, it is given as

$$\oint_C \mathbf{dx} \cdot \mathbf{E} = -\frac{\mathrm{d}}{\mathrm{d}t} \int_S \mathbf{dS} \cdot \mathbf{B} = -\frac{\mathrm{d}\Phi}{\mathrm{d}t}. \qquad (4.38)$$

Here C is the wire closed loop, S is the planar surface bounded by the wire loop and Φ is the flux of the magnetic field through the planar surface S given as $\int_s dS\ B$ $\int_s dS\ B$. The left-hand-side of Eq. (4.38), $\oint dx\ E = \varepsilon$ $\oint dx\ E = \varepsilon$ dx E = E (C), is known as the electromotive force, or EMF. The minus sign in Eq. (4.38) indicates that the current generated by the EMF would act to reduce the magnetic flux through the C loop. This sign and the fact that induced current acts to reduce any imposed magnetic field is known as Lenz's law.

A significant characteristic of Faraday's law is that when a wire is wound N times around a loop C, it captures a flux that is N times greater.

Therefore, the electromotive force (EMF) is increased by a factor of N.

$$\mathcal{E} = -N\frac{d\Phi}{dt}. \tag{4.39}$$

An inductor is a fundamental component of an electrical circuit that generates a voltage that is directly proportional to the rate at which the current flowing through it changes over time,

$$V = -L\frac{dI}{dt}, \tag{4.40}$$

The variable L represents the inductance, quantified in henrys (H), where 1 H is equivalent to 1 volt-second per ampere (V·s/A). The approximate inductance of a solenoid can be determined by considering its parameters, including the number of turns per unit length (n), the volume (V), and the core permeability (μ).

$$L_{sol} \cong n^2 \mathcal{V} \mu. \tag{4.41}$$

The equation for the energy stored in an inductor is given by EEM = $LI^2/2$. The formula can be expressed in a general form as follows:

$$E_{EM} = \frac{1}{2\pi}\int |\mathbf{B}|^2 dV, \tag{4.42}$$

The integration should be carried out across the volume of the inductor (Harland, 2000).

### 4.3.6 Electrical Devices

Electrical devices convert electrical energy into alternative forms of energy, including but not limited to light, heat, or motion. Pumps and blowers are electrical apparatuses that convert electrical energy into mechanical energy, thereby inducing fluid motion. The heater converts electrical current into thermal energy, while the electric bulb converts electrical current into visible light. Electric motors and electric generators are considered to be crucial electrical devices that are extensively utilized in various energy systems.

An electric motor is a device that converts electrical energy into mechanical energy. Electric motors find application in a diverse array of devices, encompassing cell phones, computers, blowers, pumps, elevators, and vehicular propulsion. Motors typically consist of two primary components: the rotating rotor and the stationary stator, both of which incorporate magnets.

In the most basic electric motors, the rotor consists of an electromagnet that is energized by direct current. The direction of the current flow is altered by a commutator and brushes, which are dependent on the rotor's angle. Conversely, the stator employs a permanent magnet. The theoretical average power output of a motor rotating at frequency v can be calculated using the equation N = 4nIABv, where n represents the number of windings in the wire loop of the rotor, I denotes the current flowing through the wire, A represents the area of the current loop, and B represents the constant magnetic field generated by the stator. In practical devices, it is inevitable to encounter losses resulting from friction and Joule's heating (Blaabjerg & Ionel, 2015).

### 4.3.7 Electromagnetic Energy Losses

The conversion of electromagnetic energy into thermal energy is achieved through the process of Joule heating, which is mathematically represented by equations (4.43) and (4.44). Magnetic hysteresis and eddy currents generated within electric conductors, such as steel, during the passage of time-varying magnetic fields contribute to supplementary losses. In the majority of applications pertaining to "soft" iron, eddy currents are the primary cause of energy dissipation. In order to mitigate the detrimental effects of eddy current loss, it is common practice to employ laminated magnetic circuits in transformers and electric machines. These circuits consist of thin sheets of steel, which serve to minimize the occurrence of eddy currents. In order to mitigate losses, the steel is alloyed with elements, commonly silicon, that have a detrimental effect on electrical conductivity (Maier et al., 2001).

Losses of this nature are observed across various electric devices, including motors, generators, transformers, and similar apparatus. In general, the dissipation of electromagnetic energy is significantly lower compared to thermodynamic losses, which arise from the second law of thermodynamics. In certain cases, electric devices are regarded as "ideal" machines that exhibit no losses, resulting in the output power being equivalent to the input power supplied. When the magnitude of power loss is significant, it can be determined by:

$$N_{loss} = N_{in} - N_{out}, \tag{4.43}$$

Where $N_{in}$ represents the input power that is given, while $N_{out}$ represents the output power that is obtained, for instance, in the case of an electric motor operating under steady-state conditions, the input power can be observed.

$$N_e = \sum_i V_i I_i, \tag{4.44}$$

The quantity is simply the aggregate of electrical power inputs to the distinct phase terminals. The mechanical output power can be defined as the product of torque (T) and angular velocity (ω) over a given period,

$$N_m = T\omega. \tag{4.45}$$

The calculation of electrical motor loss is subsequently determined as.

$$N_{loss} = N_e - N_m = \sum_i V_i I_i - T\omega. \tag{4.46}$$

Motor efficiency is defined as

$$\eta = \frac{N_m}{N_e} = 1 - \frac{N_{loss}}{N_e}. \tag{4.47}$$

The efficiency of electric motors is subject to variation based on their specific design and the conditions under which they are operated. Most electric motors typically exhibit efficiencies ranging from 70% to 90% (Wu et al., 2022).

### 4.3.8 Electromagnetic Energy Storage

The storage of electromagnetic energy on a large scale is a significant and complex issue. As previously demonstrated, capacitive energy storage exhibits inherent limitations and offers relatively low energy densities. The predominant technology presently employed in practical applications, yet still undergoing further advancements, pertains to rechargeable batteries for energy storage (Xu et al., 2022).

Batteries can store energy in chemical form and employ electrochemical reactions to convert energy to and from electrical form. Most residential energy storage batteries, also employed in consumer electronics, are lithium-ion based. The batteries exhibit a notable energy density, a comparatively low rate of self-discharge, and require minimal maintenance. The primary drawbacks of these devices encompass the process of aging, which occurs even in the absence of usage, the necessity of a protective circuit to ensure that voltage and current remain within acceptable thresholds, and the substantial expenses associated with their production. A lithium-ion battery consists of a cathode, an anode, and an electrolyte, which serves as the conductor. The cathode is composed of a metal oxide material, while the anode is constructed using a porous carbon material. During the discharge process, ions' movement occurs from the anode to the cathode by traversing through the electrolyte and separator. Consequently, the anode experiences oxidation. During the charging process, the direction of ion flow is reversed, resulting in the cathode acquiring electrons.

Lithium-ion batteries exhibit a wide range of variations in terms of performance and the selection of active materials. Historically, the anode material in batteries was composed of coke. However, in the 1990s, a significant shift occurred within the battery manufacturing industry, whereby graphite became the preferred choice for anode material due to its ability to achieve a more consistent voltage discharging curve. Graphene is a prospective material that holds the potential to augment the performance of lithium-ion batteries significantly (Pirtsul et al., 2022).

## SUMMARY

Mechanical and electromagnetic energy are considered two fundamental manifestations of energy within natural phenomena. Mechanical energy refers to the energy that an object possesses due to its motion and its position or configuration. Mechanical energy comprises two primary constituents, namely kinetic energy and potential energy. The principle of mechanical energy conservation asserts that within a closed system, the aggregate mechanical energy remains constant without external forces exerting work upon the system. Electromagnetic energy is a type of energy that is closely linked to the electromagnetic field. The phenomenon under consideration involves the propagation of waves characterized by a combination of electric and magnetic fields. This form of energy is accountable for many phenomena, encompassing the emission of light, radio waves, and X-rays.

## REFERENCES


1. Amores, A., Force, A., Yan, Y. L., Joly, L., Amemiya, C., Fritz, A., ... & Postlethwait, J. H. (1998). Zebrafish hox clusters and vertebrate genome evolution. Science, 282(5394), 1711-1714.
2. Belinfante, F. J. (1940). On the current and the density of the electric charge, the energy, the linear momentum and the angular momentum of arbitrary fields. Physica, 7(5), 449-474.
3. Bell, E., & Taylor, S. (2003). The elevation of work: Pastoral power and the new age work ethic. Organization, 10(2), 329-349.
4. Berger, M. A. (1985). Structure and stability of constant-alpha force-free fields. The Astrophysical Journal Supplement Series, 59, 433-444.
5. Biewener, A. A. (2006). Patterns of mechanical energy change in tetrapod gait: pendula, springs and work. Journal of Experimental Zoology Part A: Comparative Experimental Biology, 305(11), 899-911.
6. Blaabjerg, F., & Ionel, D. M. (2015). Renewable energy devices and systems–state-of-the-art technology, research and development, challenges and future trends. Electric Power Components and Systems, 43(12), 1319-1328.
7. Braams, B. J., & Bowman, J. M. (2009). Permutationally invariant potential energy surfaces in high dimensionality. International Reviews in Physical Chemistry, 28(4), 577-606.
8. Cevc, G. (1990). Membrane electrostatics. Biochimica et Biophysica Acta (BBA)-Reviews on Biomembranes, 1031(3), 311-382.
9. Cohen, W. W. (1995). Fast effective rule induction. In Machine learning proceedings 1995. Morgan Kaufmann. (Vol. 2, pp. 115-123).
10. Davis, M. H. (1964). Two charged spherical conductors in a uniform electric field: Forces and field strength. The Quarterly Journal of Mechanics and Applied Mathematics, 17(4), 499-511.
11. Gutowski, K. A., & Force, A. F. G. T. (2009). Current applications and safety of autologous fat grafts: a report of the ASPS fat graft task force. Plastic and reconstructive surgery, 124(1), 272-280.
12. Harland, R. (2000). Neural induction. Current opinion in genetics & development, 10(4), 357-362.
13. Harp, G. D., & Berne, B. J. (1968). Linear-and angular-momentum autocorrelation functions in diatomic liquids. The Journal of Chemical Physics, 49(3), 1249-1254.
14. Heinrich, R. S., Fontaine, A. A., Grimes, R. Y., Sidhaye, A., Yang, S., Moore, K. E., ... & Yoganathan, A. P. (1996). Experimental analysis of

fluid mechanical energy losses in aortic valve stenosis: importance of pressure recovery. Annals of biomedical engineering, 24, 685-694.

15. Honig, B., & Nicholls, A. (1995). Classical electrostatics in biology and chemistry. Science, 268(5214), 1144-1149.
16. Jia, Y., Li, K., Xue, L., Ren, J., Zhang, S., & Li, H. (2017). Mechanical and electromagnetic shielding performance of carbon fiber reinforced multilayered (PyC-SiC) n matrix composites. Carbon, 111, 299-308.
17. Kajita, S., Kanehiro, F., Kaneko, K., Fujiwara, K., Harada, K., Yokoi, K., & Hirukawa, H. (2003, October). Resolved momentum control: Humanoid motion planning based on the linear and angular momentum. In Proceedings 2003 IEEE/RSJ international conference on intelligent robots and systems (iros 2003) (cat. no. 03ch37453). IEEE. (Vol. 2, pp. 1644-1650).
18. Lai, Y. C., Lu, H. W., Wu, H. M., Zhang, D., Yang, J., Ma, J., ... & Dickey, M. D. (2021). Elastic multifunctional liquid–metal fibers for harvesting mechanical and electromagnetic energy and as self-powered sensors. Advanced Energy Materials, 11(18), 2100411.
19. Liwo, A., Sieradzan, A. K., Lipska, A. G., Czaplewski, C., Joung, I., Żmudzińska, W., ... & Ołdziej, S. (2019). A general method for the derivation of the functional forms of the effective energy terms in coarse-grained energy functions of polymers. III. Determination of scale-consistent backbone-local and correlation potentials in the UNRES force field and force-field calibration and validation. The Journal of Chemical Physics, 150(15), 510-515.
20. Lorenz, E. N. (1955). Available potential energy and the maintenance of the general circulation. Tellus, 7(2), 157-167.
21. Maamer, B., Boughamoura, A., El-Bab, A. M. F., Francis, L. A., & Tounsi, F. (2019). A review on design improvements and techniques for mechanical energy harvesting using piezoelectric and electromagnetic schemes. Energy Conversion and Management, 199, 111973.
22. Mahmoud, M., Ramadan, M., Olabi, A. G., Pullen, K., & Naher, S. (2020). A review of mechanical energy storage systems combined with wind and solar applications. Energy Conversion and Management, 210, 112670.
23. Maier, S. A., Brongersma, M. L., & Atwater, H. A. (2001). Electromagnetic energy transport along arrays of closely spaced metal rods as an analogue to plasmonic devices. Applied Physics Letters, 78(1), 16-18.
24. Mei, H., Zhao, Y., Han, D., Ji, T., Xu, L., Cheng, L., & Dassios, K. G. (2019). Depositing CNTs on the interface and surface of C/PyC/SiCs for tunable mechanical and electromagnetic properties. Ceramics

International, 45(17), 23411-23417.

25. Miller, D. I., & Munro, C. F. (1985). Greg Louganis' springboard takeoff: II. Linear and angular momentum considerations. Journal of Applied Biomechanics, 1(4), 288-307.
26. Munir, Z. A., Quach, D. V., & Ohyanagi, M. (2011). Electric current activation of sintering: a review of the pulsed electric current sintering process. Journal of the American Ceramic Society, 94(1), 1-19.
27. Pirtsul, A. E., Rubtsova, M. I., Mendgaziev, R. I., Cherednichenko, K. A., Kruglov, V. V., Komlev, A. S., ... & Voronin, D. V. (2022). Phase-change composites for bimodal solar/electromagnetic energy storage based on magnetite-modified cellulose microfibers. Materials Letters, 327, 132997.
28. Postlethwait, J. H., Yan, Y. L., Gates, M. A., Horne, S., Amores, A., Brownlie, A., ... & Talbot, W. S. (1998). Vertebrate genome evolution and the zebrafish gene map. Nature Genetics, 18(4), 345-349.
29. Sato, S. (1955). A new method of drawing the potential energy surface. Bulletin of the Chemical Society of Japan, 28(7), 450-453.
30. Schlegel, H. B. (2003). Exploring potential energy surfaces for chemical reactions: an overview of some practical methods. Journal of Computational Chemistry, 24(12), 1514-1527.
31. Stöhr, J., & Siegmann, H. C. (2006). Magnetism. Solid-State Sciences. Springer, Berlin, Heidelberg, 5, 236.
32. Taddeo, M., & Floridi, L. (2018). How AI can be a force for good. Science, 361(6404), 751-752.
33. Tew, J. (2006). Understanding power and powerlessness: Towards a framework for emancipatory practice in social work. Journal of Social Work, 6(1), 33-51.
34. Vanommeslaeghe, K., Hatcher, E., Acharya, C., Kundu, S., Zhong, S., Shim, J., ... & Mackerell Jr, A. D. (2010). CHARMM general force field: A force field for drug-like molecules compatible with the CHARMM all-atom additive biological force fields. Journal of computational chemistry, 31(4), 671-690.
35. Venkatachari, A. K., Halliburton, S. S., Setser, R. M., White, R. D., & Chatzimavroudis, G. P. (2007). Noninvasive quantification of fluid mechanical energy losses in the total cavopulmonary connection with magnetic resonance phase velocity mapping. Magnetic resonance imaging, 25(1), 101-109.
36. Verosub, K. L., & Roberts, A. P. (1995). Environmental magnetism: Past, present, and future. Journal of Geophysical Research: Solid Earth, 100(B2),

2175-2192.

37. Walker, D. A., Kowalczyk, B., de La Cruz, M. O., & Grzybowski, B. A. (2011). Electrostatics at the nanoscale. Nanoscale, 3(4), 1316-1344.
38. Wu, Z., Cheng, H. W., Jin, C., Yang, B., Xu, C., Pei, K., ... & Che, R. (2022). Dimensional design and core–shell engineering of nanomaterials for electromagnetic wave absorption. Advanced Materials, 34(11), 2107538.
39. Xu, P., Wang, G. C., Cai, X., Shen, H. Y., & Jiang, W. X. (2022). Design and optimization of high-efficiency meta-devices based on the equivalent circuit model and theory of electromagnetic power energy storage. Journal of Physics D: Applied Physics, 55(19), 195303.

CHAPTER

# 5 Nuclear Energy

## Contents

## INTRODUCTION

Nuclear energy has been utilized for electricity generation since 1954. Despite its initial perception as an innovative technology with the potential to provide abundant power, its historical trajectory has been marked by significant controversy. The industry has encountered numerous challenges; however, many critics of nuclear power widely regard the Chornobyl incident in 1986 as its most severe occurrence. During that period, the waning popularity of nuclear energy was attributed to a confluence of challenges and public apprehensions, further exacerbated by the growing competition posed by fossil fuels. Consequently, it appeared probable that nuclear power would be relegated to obsolescence in the annals of history. Recently, there has been a notable resurgence of interest in nuclear power, as evidenced by numerous countries constructing new plants or expressing intentions to do so. Several factors are driving the upsurge of interest. The following points will be discussed in the subsequent sections. It is important to acknowledge that the resurgence is not attributable to a singular issue but rather to various combinations exerting influence on the ongoing debate (Zhou & Zhang, 2010).

**Did You Know?**

The first recorded use of nuclear energy dates back to the 2nd century BC, where the Chinese used the radioactive materials to treat diseases.

## 5.1 FACTORS FOR SURGED INTEREST IN NUCLEAR POWER

### 5.1.1 Climate Change

This issue represents the most paramount challenge confronting the human species. The prevailing consensus in the scientific community is that human activities contribute to the intensification of the greenhouse effect, with the primary factor being the utilization of fossil fuels. Supporters of nuclear energy argue that nuclear power plants do not release greenhouse gases at the point of use. Nuclear power stations exclusively generate electrical power, in contrast to fossil fuels that yield a diverse range of energy outputs. While it is true that nuclear power plants do not emit greenhouse gases during their operational phase, it is important to acknowledge that the entire lifecycle of nuclear power, encompassing fuel production, construction, decommissioning, and waste storage, contributes to the greenhouse effect. The extent of the contribution is a matter of contention, yet it is comparatively lower than the contribution stemming from fossil fuels when considering equivalent parameters (Azam et al., 2021).

### 5.1.2 Energy Security

The human race is approaching a point referred to as the depletion of oil resources. Fossil fuels possess a limited lifespan, and the concept of 'peak oil', denoting the point at which consumption surpasses production, is believed to be imminent or potentially occurring. The industrialized nations heavily rely on fossil fuels, and a significant portion of these energy sources are in regions characterized by political instability (Bang, 2010). The assurance of supply security represents a significant challenge. Supporters of nuclear power contend that its utilization has the potential to bolster energy security due to the abundant availability of nuclear fuel and the implementation of reprocessing techniques that can effectively extend the lifespan of depleted fuel. Some individuals hold differing viewpoints and contend that fast breeder technology can generate fuel. Considering these factors collectively, it can be argued that an increase in nuclear capacity would reduce dependence on imported resources and enhance the diversification of the energy mix (Hayashi & Hughes, 2013).

### 5.1.3 Costs

The historical record of nuclear power reveals a recurring pattern of cost underestimation, highlighting the presence of lingering uncertainties surrounding the accurate assessment of long-term storage expenses associated with nuclear waste, particularly high-level waste. When comparing nuclear power generation to fossil fuels, it can be observed that the former has exhibited a slightly higher cost. There has been a recent shift in this phenomenon due to the substantial increase in fossil fuel costs. Proponents assert that as nuclear build programs gain more experience and adopt greater standardization, the management of costs will improve. Additionally, they argue that nuclear power becomes a cost-effective alternative when compared with fossil fuel generation as the role of fuel in operating costs diminishes (Koomey & Hultman, 2007).

### 5.1.4 Safety

Supporters of nuclear power contend that there has been a significant improvement in safety measures since the Chornobyl incident. Furthermore, they assert that the increasing global utilization of nuclear energy will contribute to developing a stronger safety culture within the nuclear industry. The inherent safety of the new generation of nuclear reactors is a subject of debate, with proponents asserting that it surpasses that of earlier designs, further augmenting safety measures. Nevertheless, concerns persist regarding the proper handling and elimination of nuclear waste.

In conclusion, despite the established status of nuclear energy as a technological advancement, it remains a subject of significant controversy. This situation will probably persist (Rabbi & Sabharwal, 2018).

## 5.2 TECHNOLOGY OVERVIEW

A nuclear power facility can be characterized as a conventional means of electricity generation. The operational principles of this system closely resemble those of a conventional coal or gas-fired power plant. The reactor produces thermal energy that is utilized to generate steam, subsequently propelling a conventional steam turbine. In essence, this process involves the boiling of water. Similar to a coal or gas-fired power plant, the system's overall efficiency is determined by the principles outlined in the principles of thermodynamics. One notable distinction between the technologies is how heat is produced within a nuclear reactor. Coal and gas-fired power plants produce thermal energy through the process of combustion. The combustion process occurs when a fuel substance undergoes a chemical reaction with oxygen, resulting in thermal energy generation. Distinct mechanisms characterize the generation of energy or heat in a nuclear reaction. Nuclear energy is generated through two distinct methods (Zinkle et al., 2002):

- Fission: Fission refers to the process in which a heavy nucleus undergoes division, forming several radioactive nuclei. The release of gamma rays, neutrons, and a substantial quantity of energy accompanies this division. Fission is typically initiated through the absorption of a neutron by a heavy nucleus, although it can also occur spontaneously.
- Fusion: Fusion is a phenomenon characterized by merging two atomic nuclei, resulting in the liberation of a substantial amount of energy. Fusion processes power the sun (Turconi et al., 2013).

All nuclear reactors employ fission to generate energy, which is subsequently utilized to produce electricity by raising steam to power a turbine. Developing a fusion reactor has garnered significant attention and is an area of active investigation. However, it is anticipated that the realization of a functional fusion reactor will require a substantial amount of time, potentially spanning several decades, to reach its completion (Kang et al., 2009).

**Did You Know?**

On December 20, 1951, EBR-1 in Idaho, USA, became the world's first electricity generating nuclear power plant. It produced enough electricity to light four 200-watt light bulbs.

### 5.2.1 Nuclear Reactions

Atoms consist of a central nucleus, which is surrounded by electrons. The nucleus can accommodate both protons and neutrons. The particles above are commonly referred to as nucleons. Protons possess a positive charge, while neutrons are electrically neutral, lacking any charge. Electrons, on the other hand, exhibit a negative charge. The mass of a single atom is determined by the total number of neutrons and protons it possesses. Various elements are characterized by their atomic number, denoting the number of protons housed within the nucleus. Naturally occurring elements exhibit a wide range of mass, from the least massive element, hydrogen, composed of a single proton and electron, to Uranium, possessing an atomic number of 92. Like other chemical elements, Uranium exhibits variations in its atomic structure, giving rise to distinct variants referred to as isotopes. Isotopes of a given element exhibit variations in the number of neutrons within the nucleus. Natural Uranium is primarily composed of a combination of two isotopes, namely uranium-238 (U-238) and uranium-235 (U-235). Uranium-238 consists of 92 protons, 92 electrons, and 143 neutrons, while uranium-235 contains 92 protons, 92 electrons, and 146 neutrons. Uranium-238 comprises approximately 99.3 percent of the total uranium composition, while uranium-235 constitutes approximately 0.7 percent (Rosa & Freudenberg, 1993).

### 5.2.2 Nuclear forces

In the realm of natural phenomena, there exist four fundamental forces, namely the electromagnetic force, the strong nuclear force, the weak nuclear force, and gravity. The initial trio of forces plays integral roles in the constitution of the atomic nucleus. The subatomic particles constituting nucleons are held together by the powerful force known as the strong nuclear force. In science, the Standard Model symbolizes a nucleon encompassing 16 sub-atomic particles. These particles can be further categorized into 12 particles of matter and 3 force-carrier fragments. The residual strong force, responsible for the binding of a neutron and a proton, is generated by the influence of force-carrier particles near the nucleon's boundary. Due to an electromagnetic field, two protons tend to repel one another. The cohesive nature of the atomic nucleus is primarily attributed to the presence of neutrons and a powerful residual force (Tews et al., 2020).

Due to their lack of charge, neutrons do not contribute to the preexisting repulsion between protons, thereby aiding their spatial separation. This phenomenon mitigates the intense electrostatic repulsion exerted by adjacent protons. Neutrons contribute significantly to the strong residual force within the nucleus. The equilibrium of forces contributes to the maintenance of stability within the nucleus. As the size of the nucleus expands, there is a corresponding increase in the number of nucleons, resulting in an amplified repulsive force. It is necessary to augment the intensity of the powerful forces to achieve equilibrium (Freedman, 1981). As the number of protons increases, the corresponding number of neutrons also increases, resulting in a gradual rise in the percentage of neutrons to protons. This ratio exceeds 1.5 for the largest nucleus. Ultimately, a threshold is attained at which the nucleus undergoes instability. The nucleus of bismuth, consisting of 83 protons and 126 neutrons, is recognized as the most substantial stable nucleus. Nuclei containing a proton count exceeding 83 are universally recognized as inherently unstable, leading to their eventual decay through a process commonly referred to as radioactivity. It is important to note that Uranium possesses a total of 92 protons (Beane et al., 2002).

### 5.2.3 Decay

The decay of an unstable nucleus occurs through the emission of a particle, transforming the nucleus into either a different nucleus or a state with less energy. A series of decay events occur until a nucleus achieves stability. Three prevalent forms of radioactive decay exist alpha, beta, and gamma. The distinction between these entities lies in the type of particle released by the nucleus as part of the decay process (Altmann et al., 1995).

Alpha decay is a radioactive decay process that transpires when the atomic nucleus possesses excess protons. This phenomenon leads to an excessive repulsive force, resulting in the emission of a helium nucleus to mitigate the repulsion. Alpha particles have a limited range of air travel due to their absorption propensity (Srivastava & Kulkarni, 2015).

Beta decay is a nuclear process that takes place when the ratio of neutrons to protons within the nucleus exceeds a certain threshold. A neutron transforms into a proton and an electron in the beta-decay process. Subsequently, the electron is discharged. Positron emission can occur when the neutron-to-proton ratio is insufficiently balanced. During a process known as beta-plus decay, a proton transforms a neutron and a positron, subsequently releasing the positron.

Gamma decay is a phenomenon that arises due to excessive energy levels within the nucleus. The nucleus transitions to a lower energy level, resulting in the emission of a high-energy photon referred to as a gamma particle. Gamma emissions possess high penetrating power. However, their absorption can be maximized by employing a considerably thick coating of high-density material, such as lead (Stohl et al., 2012).

### 5.2.4 Nuclear Energy

The combustion of fossil fuels involves the chemical reaction between hydrocarbons present in these fuels and oxygen in the atmosphere, resulting in water and carbon dioxide formation. The phenomenon above refers to a chemical reaction wherein the valence electrons of atoms participate, resulting in the liberation of energy via the reconfiguration of chemical bonds. Nuclear energy arises from the reconfiguration of the constituents within the atomic nucleus. Due to the significantly stronger interatomic forces exhibited by the inner constituents compared to the outer electrons, nuclear reactions yield a substantially higher amount of energy. Nuclear reactions tend to release approximately one hundred million times more power than chemical reactions. This particular reaction powers the nuclear reactor (Petrescu et al., 2016).

The nucleus is comprised of subatomic particles known as protons and neutrons. However, it is noteworthy that the total mass of a nucleus is consistently lower than the combined masses of its constituent protons and neutrons. The disparity is quantified as the nuclear binding energy, which serves as the cohesive force maintaining the integrity of the nucleus. The phenomenon characterized by a mass discrepancy is commonly called a mass defect. When a material with a high mass, such as Uranium, undergoes fission, it forms two nuclei with lower mass and increased stability. This process is accompanied by a reduction in mass, which is released as kinetic and radiation. The quantification of energy provided can be ascertained through the utilization of Einstein's equation, which establishes a correlation between energy and mass (Koning & Rochman, 2008):

$$\ddot{u} = \quad {}^{2}$$

In the given equation, E represents energy, m represents mass, and c represents the speed of light.

The mass defect of a nucleus refers to the disparity in mass between its constituent parts when they are isolated and the overall mass of the nucleus itself. If Δm represents the mass defect,

$$m \aleph (A - Z)m_n \quad Zm_p \quad M$$

Let M denote the total mass of the nucleus, mn represent the mass of a neutron, and mp indicate the mass of a proton, all expressed in mass units (Lenzen, 2008).

When the equation *E=mc2* is employed to transform the mass defect into an equivalent quantity of energy, the resulting energy corresponds to the binding energy (BE) of the nucleus and can be expressed as follows (Apergis et al., 2010):

$$BE(\text{Joules}) = mc^2$$

$$= \grave{\imath}\left[¢A - Z)m_n + Zm_p - M\right] \quad {}^{2}$$

(ì is the unified mass unit = $1.66 \times 10^{-27}$ kg). Or, it can be expressed in MeV using

$$BE(\text{MeV}) = 931\left[(A - Z)m_n + Zm_p - M\right]$$

The measured binding energies of various nuclei have been determined (Was et al., 2019). The findings are depicted in curves. The curve exhibits a relatively shallow peak within the mass number range of approximately 50 to 110. This implies that these nuclei exhibit the highest degree of stability. Additionally, it signifies the existence of two potential mechanisms for liberating energy from the atomic nucleus. In the preceding discussion, energy generation through nuclear fission was explored, wherein the formation of stable atomic nuclei leads to mass conversion into energy. This phenomenon arises when substances with a significantly elevated atomic mass are employed. A positive correlation between the mass and the yield indicates that as the mass increases, the yield also increases (Kunze et al., 2010). The graph's left-hand side illustrates that the binding energy curve exhibits a significantly greater slope for nuclei characterized by lower mass numbers. Combining two light nuclei to create a nucleus results in the liberation of significantly higher amounts of energy than splitting a heavy nucleus. While the fusion reaction has been successfully demonstrated in the context of thermonuclear or hydrogen bombs, the controlled utilization of fusion for power generation remains an ongoing challenge. The discussion of progress achieved thus far will be deferred later. As previously mentioned, nuclear reactions' energy yield far surpasses combustion reactions by several orders of magnitude. By utilizing relevant equations, it is possible to compute the energy generated from a fission reaction.

Based on research, it is evident that a nucleus possessing a mass number of 220 exhibits nucleons with an average binding energy of 7.8MeV. Conversely, a nucleus with a mass number of 110 demonstrates an average binding energy per nucleon of 8.6MeV.

When the heavy nucleus undergoes fission, resulting in the production of two lighter nuclei, there is an increase in the average binding energy of the nucleons and a subsequent release of energy:

$$(8.6 - 7.8) \times 220 = 176\text{MeV}$$

The precise quantity of energy liberated, primarily attributed to the kinetic energy of the two smaller nuclei, is contingent upon the fissioned nucleus's mass and the resulting nuclei's masses. A commonly employed estimate for the energy released during the fissioning of a substantial material is approximately 200 MeV per fission. The energy quantity mentioned is commensurate with a reduction in mass of approximately 0.2 units per fission event (Bowman et al., 1992).

The quantity of U-235 utilized during the continuous operation of the power station for a year amounts to 1158kg, approximately equivalent to 1.2 metric tonnes. In comparison, a coal-fired power station with a comparable level of electricity generation would typically require the combustion of approximately 2.5 million metric tonnes of coal, as Hill et al. (1995) indicated. A conventional nuclear reactor with a capacity of 1000 megawatts (MWe) has the potential to generate sufficient electrical power to meet the demands of a contemporary urban area accommodating approximately one million individuals (Akyüz, 2017).

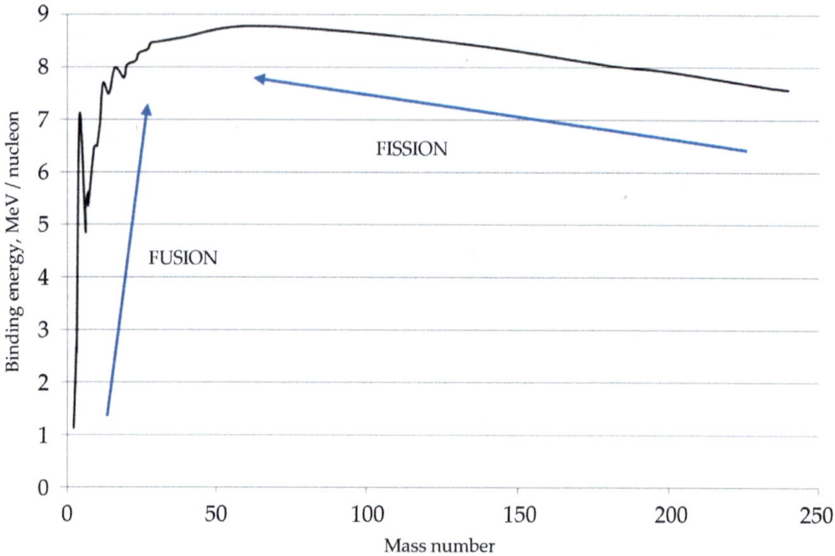


**Figure 5.1.** Illustration of the graph representing binding energies (Source: Tosti S. Classical Thermodynamic Analysis of D-Based Nuclear Fusion Reactions: The Role of Entropy. Energies. 2023; 16(10):3996. https://doi.org/10.3390/en16103996, Creative Commons License).

**Remember**

Nuclear energy has been utilized in space missions to power satellites, probes, and even rovers.

## 5.2.5 Nuclear Reactions for a Nuclear Power Station

U-258, an inherently unstable atomic nucleus, necessitates specific environmental circumstances to undergo fission and subsequently emit energy. When a U-235 atom's nucleus undergoes a process of neutron capture, it undergoes nuclear fission, dividing the nucleus into two fragments. This fission process releases thermal energy, accompanied by the emission of two or three additional neutrons. If a sufficient number of these ejected neutrons induce the division of the nuclei of other U-235 atoms, resulting in the liberation of additional neutrons, a process known as nuclear fission can lead to the establishment of an autonomous sequence of reactions referred to as a 'chain reaction'. When this phenomenon occurs repeatedly, spanning a significant magnitude, an extensive quantity of thermal energy is generated from a comparatively limited quantity of Uranium. The significance of the neutron here is noteworthy (Teller et al., 1996).

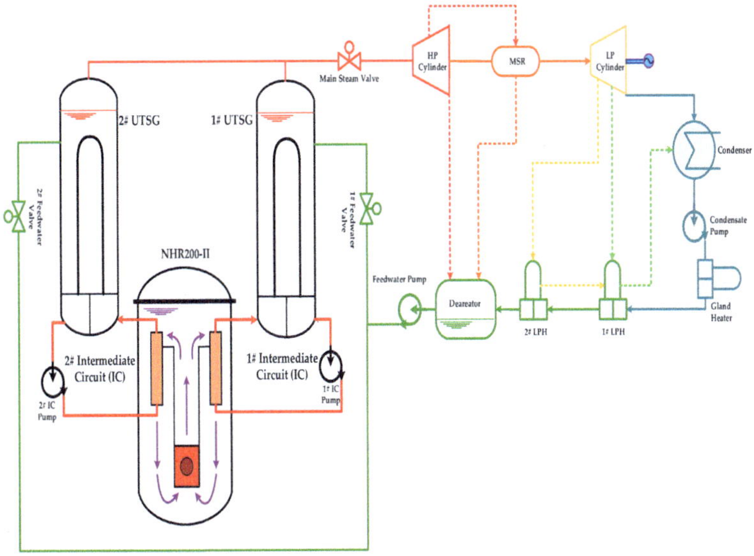


**Figure 5.2.** Illustration of a nuclear power plant (Source: Dong Z, Liu M, Jiang D, Huang X, Zhang Y, Zhang Z. Automatic Generation Control of Nuclear Heating Reactor Power Plants. Energies. 2018; 11(10):2782. https://doi.org/10.3390/en11102782, Creative Commons License).

In the context of nuclear reactions, the absence of electric charge in neutrons eliminates the repulsion issue. Consequently, neutrons are not subject to the repulsive force and can more readily interact with atomic nuclei. Fission reactions and other significant processes that necessitate comprehension of nuclear reactors entail the interplay between neutrons and atomic nuclei. The fundamental requirement within a reactor is to establish an environment that facilitates the provision of neutrons possessing an optimal kinetic energy, thereby enabling the fission of U-235 to occur at a desirable pace. Additionally, it is imperative to ensure the ability to regulate the reactor by adjusting the neutron availability as required (Saouma & Hariri-Ardebili, 2014).

When a neutron collides with U-235, it produces several fission products along with a certain number of neutrons, typically averaging 2.4 neutrons per reaction. Assuming the presence of an adequate number of U-235 atoms nearby, the emitted neutrons will engage in interactions with these atoms, resulting in the production of additional fission products and neutrons. The process above represents the fundamental chain reaction inherent in nuclear power, which facilitates the generation of energy utilized in producing steam for electricity generation (Sunal & Sunal, 1999). The completion of chain reactions would be unattainable if, on average, one or fewer fission neutrons were generated per fission event, as a certain portion of these neutrons would inevitably fail to undergo interactions with additional U-235 nuclei. The neutrons produced by fission can participate in various types of neutron reactions. In general, these neutrons are engaged in three distinct processes. Neutrons can undergo three possible outcomes upon interacting with fissile material: they may either leak out of the fissile material and escape, be absorbed by the fissile material or surrounding material leading to non-fission reactions, or be absorbed by U-235 and induce additional fission reactions. If, on average, a single neutron resulting from fission leads to additional fission, the reaction can be considered autonomous. In cases where multiple neutrons result in further fissions, fissions will progressively increase over time (Johnson & Zeigler, 1988).

Conversely, if less than one neutron leads to additional fission, the number of fissions will gradually decline over time. The three possibilities above are delineated by a variable denoted as k, which represents the reaction multiplication constant. If k is greater than 1, the reaction can be classified as supercritical, resulting in a progressive increase in the number of fissions over time. Conversely, if k equals 1, the reaction is critical, leading to a constant reaction rate. Lastly, if k is less than 1, the reaction is categorized as sub-critical, causing a decline in fissions as time progresses. To achieve operational functionality and control in a nuclear reactor, it is imperative to establish the capability to initiate a self-sustaining chain reaction (Freudenburg & Baxter, 1985).

### 5.2.6 The Reactor Core

Several criteria must be met to construct a stable nuclear reactor. The following items are as follows:

- The core must possess an adequate amount of fissile material to sustain a chain reaction of the necessary magnitude. The outcome is contingent upon various factors, including the quantity of U-235 present, the arrangement of the core, as well as the characteristics of the moderator and coolant substances utilized (as described subsequently). Enriched Uranium is a variant in which the concentration of U-235 is elevated to approximately

3 percent. Further discussion on reactor types will be presented in subsequent sections of this chapter. During the lifespan of a fuel load, a portion of the neutrons undergo absorption by U-238, resulting in plutonium reactions. Uranium-238 (U238) has the potential to undergo occasional fission (Durmayaz & Yavuz, 2001).

- For a fission event to occur, it is necessary for at least one neutron produced during the event to induce another fission event. The outcome is contingent upon the variables outlined in the first factor and the profile, or reaction cross-section, of U-235. It is necessary to decelerate the energetic fission neutrons to increase the likelihood of inducing further fissions.
- It is imperative to ensure that an uncontrolled supercritical chain reaction remains unattainable. Control rods are composed of materials that exhibit a high degree of neutron absorption. The function of these rods is to limit the number of neutrons within the core to the necessary operational thresholds. In emergencies, supplementary control rods are implemented. This application's appropriate materials include steel incorporating dispersed cadmium, boron, or indium (Davydov et al., 2019).
- Heat dissipation is necessary to prevent the core from experiencing excessive temperatures. It is imperative to utilize a coolant that is compatible with the moderator. In Magnox and Advanced Gas-cooled Reactors (AGRs), carbon dioxide gas serves as the coolant, while in Pressurized Water Reactors (PWRs), water functions as both the coolant and the moderator. The coolant undergoes circulation within a heat exchanger, generating steam that is utilized to drive the turbines.
- It is imperative to ensure the containment of radiation. Containment is implemented at multiple levels. The fuel is contained within cylindrical tubes composed of suitable metals such as magnox, zircalloy, or stainless steel to prevent direct contact between the fission products and the coolant. The core and coolant are enclosed within a hermetically sealed system, and this entire system is enveloped by a layer of reinforced concrete, which functions as a protective barrier against biological hazards. Various radiation detectors are strategically positioned at critical components of the system to detect potential leaks (West et al., 2011).

### 5.2.7 Reactor Technology

It is crucial to acknowledge that nuclear power stations are distinct from traditional electrical stations solely due to their utilization of heat generated by a nuclear reactor to produce steam for their turbines.

The foundation of any reactor design is determined by several key factors, including the choice of fuel, the moderator employed, the control systems implemented, and the medium utilized for heat transfer from the reactor. There exist two fundamental methodologies. The first approach involves utilizing distinct materials for the moderator and coolants, exemplified by Magnox and Advanced Gas Cooled reactors (AGR). The second option employs a common substance as a moderator and a coolant, such as the Pressurized Water Reactor (PWR) (Rahmanta et al., 2023).

## 5.3 MAGNOX AND AGR REACTORS

The Magnox reactor type, which derived its name from utilizing a magnesium alloy for constructing the fuel can housing the uranium fuel, emerged as one of the initial commercial nuclear stations. The Magnox reactors employed natural uranium metal as the primary fuel source, incorporated a graphite moderator, and utilized pressurized carbon dioxide as the coolant medium. These entities are currently considered outdated or no longer in use. In the United Kingdom, AGR has emerged as the successor (refer to Figure 5.3). The Advanced Gas-cooled Reactor (AGR) employs stainless steel cans clad with enriched Uranium, a graphite moderator, and pressurized carbon dioxide (CO2) as the coolant. Advanced gas-cooled reactors (AGRs) are characterized by their ability to operate at elevated temperatures compared to the Magnox reactor. The Advanced Gas-cooled Reactor (AGR) is housed within a robust pre-stressed concrete pressure vessel, reinforced with steel lining, with a substantial several-meter thickness. This vessel functions as a biological shield, effectively containing the reactor and accommodating the boilers within its structure. The coolant facilitates the transfer of thermal energy from the reactor to the boilers, which raises the water temperature within a closed steam circuit. The generated steam is subsequently utilized to drive the turbines, similar to the operation of coal, oil, or gas-fired power plants (Standring, 1966).

Another design that employs distinct materials for the moderator and coolant is the RBMK (Reactor Bolshoy Moshchnosty Kanalny), a high-power channel reactor developed by the Soviet Union.

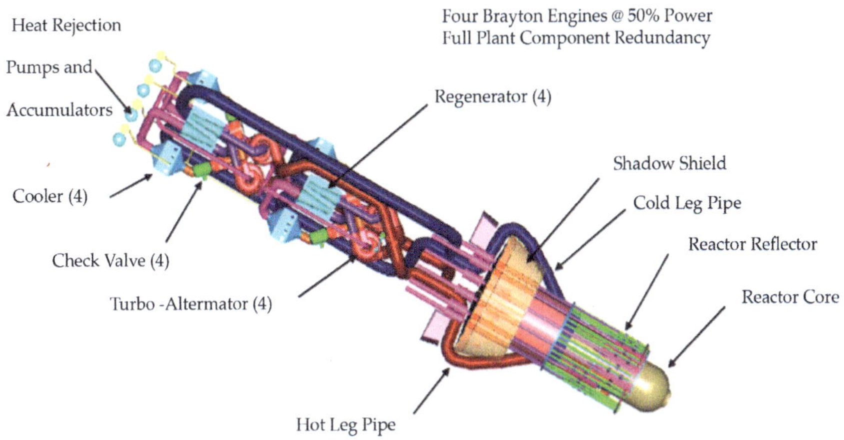


**Figure 5.3.** Diagram of an AGR nuclear reactor (Source: *Čížek* J, Kalivodová J, Janeček M, Stráský J, Srba O, Macková A. Advanced Structural Materials for Gas-Cooled Fast Reactors—A Review. Metals. 2021; 11(1):76. https://doi.org/10.3390/met11010076, Creative Commons License).

The pressurized water reactor (PWR) employs regular water as its cooling fluid and graphite as its moderator. This particular power reactor design distinguishes itself from most other designs due to its original purpose and utilization for plutonium production and power generation (Metcalfe & Mills, 2015).

## 5.4 WATER REACTORS

Water-cooled reactors represent the prevailing nuclear design worldwide, using a single material for both moderator and coolant functions. There exist various classifications, which can be broadly categorized into two primary groups.

Heavy water reactors utilize heavy water as both the moderator and coolant. A water molecule is composed of two hydrogen atoms and one oxygen atom. Most water consists of the hydrogen/oxygen compound, while a minor proportion contains an additional hydrogen isotope, deuterium, and oxygen. Deuterium is distinguished from hydrogen due to the presence of a single neutron within the nucleus of each atom. This substance is commonly referred to as heavy water. The utilization of heavy water, which contains deuterium, exhibits a modestly heightened efficacy in the deceleration of neutrons originating from fission reactions. Consequently, this characteristic enables the utilization of natural Uranium as a viable fuel source. The reactors of this particular type are commonly referred to as CANDU reactors in Canada (Corradini et al., 1988).

Light water reactors (LWRs) are nuclear reactors that utilize ordinary water (H2O) as both the coolant and the moderator. There exist two distinct variants of the light water reactor. One type of reactor is known as the boiling water reactor (BWR). In this particular design, the water that traverses the reactor core serves the dual purpose of moderating and cooling the reactor while also serving as the primary source of steam for the turbine.

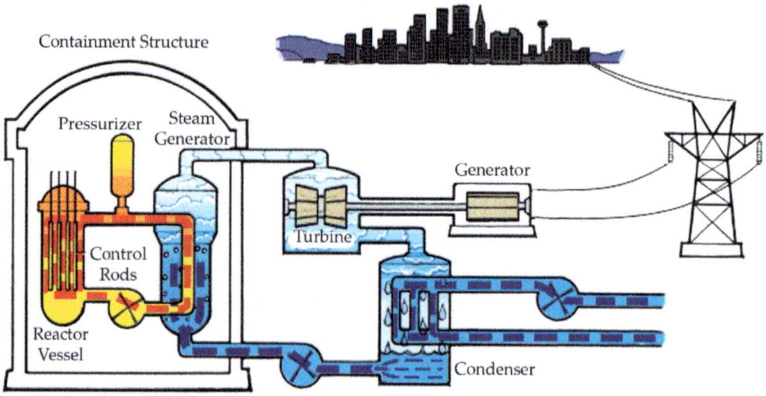


**Figure 5.4.** Diagram of a pressurized water reactor (Source: By U.S.NRC. - http://www.nrc.gov/reading-rm/basic-ref/students/animated-pwr.html, Public Domain, https://commons.wikimedia.org/w/index.php?curid=2320214, Creative Commons License).

The Pressurized Water Reactor (PWR) design is predominantly derived from United States technological advancements and remains the prevailing type of reactor employed globally (see Figure 5.4). The reactor is enclosed within a pressure vessel constructed of steel. The reactor utilizes pressurized water as a dual-purpose medium, serving as a moderator and a coolant. This water is circulated throughout the system, including the reactor and the boilers, through pumping. The primary pressurized circuit consists of the pressure vessel, boilers, and connecting pipework, all enclosed within a containment building made of pre-stressed concrete and lined with steel (Elsheikh, 2013).

## 5.5 FAST BREEDER REACTORS (FBR)

Under optimal operational circumstances, the neutrons emitted by fission reactions can facilitate the production of additional fuel from isotopes that are typically incapable of undergoing fission. The prevalent reproductive process involves the conversion of non-fissile uranium-238 into plutonium-239. The term "fast breeder" pertains to configurations that can generate a greater amount of fissile fuel than the quantity they consume. An example of such a configuration is the Liquid Metal Fast Breeder (LMFBR). The plausibility of this situation arises from the significant disparity in abundance between non-fissile uranium-238 and fissile U-235, with the former being approximately 140 times more prevalent. Moreover, the neutrons emitted during a fission chain reaction can effectively facilitate the conversion of uranium-238 into plutonium-239 (Kambe & Uotani, 1998).

Its original purpose was to enhance the global uranium reserves, potentially multiplying them by a factor of approximately 60. When the availability of those resources was deemed limited, multiple countries initiated comprehensive Fast Breeder Reactor (FBR) development initiatives. Nevertheless, the project faced notable technical and materials challenges, and the findings of geological exploration in the 1970s indicated that scarcity concerns would not be imminent. Due to the convergence of these two factors, it became evident during the 1980s that fast breeder reactors (FBRs) would not possess commercial competitiveness compared to the already established light water reactors. Despite advancements in technology, the economic viability of Fast Breeder Reactors (FBRs) remains contingent upon the relative value of the bred plutonium fuel compared to the cost of fresh Uranium.

Furthermore, there exists a global apprehension regarding the management of decommissioned military plutonium, leading to suggestions of employing fast reactors as a potential solution (WNA, 2008d). The significance of technology in global energy sustainability is evident in both aspects discussed. Consequently, there have been appeals for a more coordinated approach toward Fast Neutron Reactor (FNR) technology and the related concerns surrounding reprocessing (Kobayashi et al., 1998).

## 5.6 NEW FISSION TECHNOLOGIES

Research and development on third-generation reactors has been ongoing since 1996. The implementation of a standard design for each specific type characterizes third-generation reactors.

- This standardization aims to facilitate the licensing process, minimize capital expenses, and decrease the duration of construction.

- The design of these devices is characterized by their simplicity and durability, which enhances ease of operation and reduces susceptibility to operational disruptions (Abdulla et al., 2017).
- It is common for greater availability and operating life to span approximately 60 years.
- Decreased likelihood of core melt accidents.
- Enhanced ability to withstand significant damage that could result in releasing radioactive materials from an aircraft collision.
- Increasing the level can effectively decrease fuel consumption and minimize waste generation. Additionally, using burnable absorbers, commonly referred to as "poisons," can prolong the lifespan of fuel.

One notable deviation from second-generation designs is the inclusion of passive or inherent safety features, which do not necessitate active controls or functional intervention to prevent incidents in the event of a malfunction. These features may rely on natural forces such as gravity, natural convection, or resistance to high temperatures (Chadwick et al., 2011).

Most of the present generation comprises evolutionary designs, such as the advanced boiling water reactor (ABWR), a derivative of a General Electric design that incorporates knowledge gained from previous light water reactors (LWRs). In Europe, various designs are developed to fulfill the European Utility Requirements (EUR) set by French and German utilities, which impose rigorous safety standards. The European Pressurized Water Reactor (EPR) was developed to establish a new benchmark for nuclear power generation in France. It obtained design approval in the year 2004. The initial EPR unit is currently under construction in Olkiluoto, Finland, while the second unit is being built in Flamanville, France. Nevertheless, recent investigations have revealed many deficiencies in designing and implementing these nuclear reactors (Chadwick et al., 2006).

Except for the PMBR (pebble bed modular reactor), all other reactors employ either heavy or light water as the moderator and coolant. The Prime Minister's Briefing Room (PMBR) employs a nuclear reactor with graphite moderation and gas cooling. The Generation IV International Forum (GIF) was established in 2000 by a consortium of international nations with a substantial interest in nuclear power. The primary objective of the GIF, which was officially established in 2001, is to engage in collaborative efforts to advance the development of future nuclear technologies (Kuteev & Goncharov, 2020).

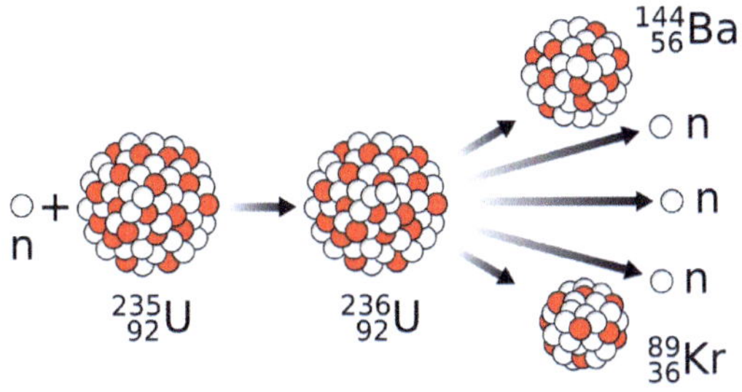


**Figure 5.5.** Illustration of a nuclear fission process (Source: By MikeRun - Own work, CC BY-SA 4.0, https://commons.wikimedia.org/w/index.php?curid=60375907, Creative Commons License).

## 5.7 FUSION

Fusion refers to amalgamating light atoms to form a more massive atom. The reaction, as mentioned earlier, yields a substantial quantity of energy. Furthermore, achieving fusion in light atoms necessitates a substantial expenditure of energy. The factor above has been a primary limitation for developing fusion energy. To initiate the energy release, fusion fuel, consisting of various hydrogen isotopes, necessitates exposure to exceedingly high temperatures exceeding 10 million degrees Celsius (Plompen et al., 2020).

The issue at hand pertains to the development of a technique for elevating the temperature of the fuel to a sufficiently high level and sustaining its confinement for a duration that enables the release of a greater amount of energy through fusion reactions than the energy expended to initiate the reaction. There are ongoing investigations into two distinct experimental methodologies: magnetic confinement fusion (MFE) and inertial confinement fusion (ICF). The initial approach involves using robust magnetic fields to confine the high-temperature plasma. The second method entails the compression of a hydrogen pellet by applying intense lasers or particle beams, resulting in its consolidation (Entler et al., 2018).

Magnetic confinement is achieved through the utilization of magnetic fields. The fuel undergoes a process of plasma heating. The transition of a gas into a plasma occurs when the application of heat or alternative forms of energy induces a substantial portion of atoms to relinquish some or all of their electrons. The residual components of these atoms retain a positive charge, while the dissociated negative electrons exhibit mobility. The atoms undergo ionization, resulting in the formation of an electrically charged gas. A plasma is formed when a sufficient number of atoms undergo ionization, resulting in a notable impact on the electrical properties of the gas (Yi-can, 2001). The utilization of magnetic confinement relies on the electrical properties of the plasma. The objective is to mitigate contact with any tangible surfaces that may result in a decrease in the temperature of the plasma. An increased amount of energy is necessary to elevate the temperature of the plasma to approximately 10 million degrees Celsius (Çakar et al., 2022).

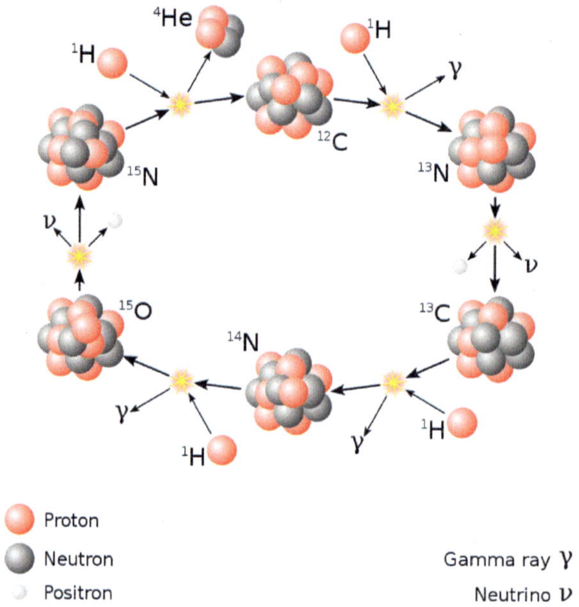


**Figure 5.6.** Illustration of a nuclear fusion reaction (Source: By Borb - Own work based on: Fusion in the Sun.svg:, Public Domain, https://commons.wikimedia.org/w/index.php?curid=691758, Creative Commons License).

The tokamak has emerged as the most promising design for achieving magnetic confinement. The Joint European Torus (JET) currently holds the distinction of being the largest operational tokamak globally. The facility is situated in Culham, Oxfordshire, United Kingdom. Experiments have been conducted at the location above since the year 1983. Thus far, the device has successfully attained a fusion power output of up to 16MW for one second in D-T plasmas (Bloch et al., 2001). The Japan Experimental Tokamak (JET) facility conducts many experiments to investigate various heating schemes and other techniques. The Joint European Torus (JET) has demonstrated significant success in effectively implementing remote handling techniques within a radioactive environment to modify the device's interior. This achievement proves that remote handling maintenance of fusion devices is indeed feasible (Dobran, 2012).

In 2006, a collaborative effort was established among the European Union, the United States, Russia, Japan, South Korea, and China to embark on a joint research initiative to explore nuclear fusion. The construction of the International Thermonuclear Experimental Reactor (ITER) project is currently underway at the Cadarache facility located in France. The projected duration of the program is expected to span 30 years, with 10 years allocated for the construction phase and 20 years designated for the operational phase. The estimated cost of the program is approximately €10 billion. The ITER project aims to generate a continuous fusion power output of around 500MW for up to 1000 seconds (Toschi, 1997).

In comparison, the JET project achieved a maximum power output of 16MW for less than one second. This power is generated through the fusion of approximately 0.5g of a deuterium/tritium (D-T) mixture within the reactor chamber, which has a volume of approximately 840m3. While it is anticipated that ITER will generate a significantly greater amount of energy (in the form of heat) compared to the energy consumed to raise the plasma to fusion temperatures, it should be noted that the resulting heat will not be utilized for electricity generation purposes (Horlick-Jones et al., 2012).

Inertial confinement (ICF) represents a relatively recent area of investigation. This method involves the precise focusing of laser or ion beams onto the surface of a target, specifically a spherical D-T ice target with a diameter of a few millimeters. The process described involves the evaporation or ionization of the material's outer layer, forming a plasma crown. This crown subsequently expands, leading to the generation of an inward-moving compression front or implosion. As a consequence of this implosion, the inner layers of the material experience an increase in temperature (Bethe, 1979). The central region of the fuel, commonly referred to as the core or central hot spot, has the potential to undergo compression up to a magnitude of one thousand times its liquid density. The ignition process is initiated once the core temperature reaches the capsule's threshold. The fuel's inertia constrains the duration of these reactions, hence the terminology, yet remains below one microsecond. The objective is to generate recurrent micro-explosions (Bock et al., 1982).

## SUMMARY

Despite the contentious nature of nuclear power, a significant level of interest exists in expanding its capacity, primarily motivated by concerns related to climate change and the security of energy supply. This assertion holds particular validity within the Organisation for Economic Co-operation and Development (OECD) context, where member countries generally exhibit a scarcity of fossil fuel resources. Within the European Union (EU), an observable trend indicates a growing inclination towards nuclear power, as evidenced by public opinion. However, it is important to note that perspectives on nuclear power remain diverse and varied on a global scale. The field of nuclear technology has witnessed advancements in safety measures following the Chornobyl incident. However, it is important to acknowledge that, as with any intricate technological system, the possibility of accidents or failures cannot be eliminated. Implementing adequate safeguards and containment measures within nuclear systems significantly reduces the probability of a recurrence of an incident similar to the Chornobyl disaster. However, there exist certain challenging matters that persist. The issue of finding a durable and secure approach to managing waste, whether from decommissioning processes or spent fuel, poses significant challenges primarily due to the extended time frames involved. The current state of our technological capabilities restricts our ability to contemplate events and developments that extend beyond a limited temporal horizon. Despite its reliability, nuclear power is expected to continue generating controversy. The prospects of fusion remain uncertain.

## REFERENCES

1. Abdulla, A., Ford, M. J., Morgan, M. G., & Victor, D. G. (2017). A retrospective analysis of funding and focus in US advanced fission innovation. Environmental Research Letters, 12(8), 084016.
2. Akyüz, E. (2017). Advantages and disadvantages of nuclear energy in Turkey: Public perception. Eurasian Journal of Environmental Research, 1(1), 1-11.
3. Altmann, M., Declais, Y., Feilitzsch, F. V., Hagner, C., Kajfasz, E., & Oberauer, L. (1995). Search for the electron-positron decay of axions and axionlike particles at a nuclear power reactor at Bugey. Zeitschrift für Physik C Particles and Fields, 68, 221-227.
4. Apergis, N., Payne, J. E., Menyah, K., & Wolde-Rufael, Y. (2010). On the causal dynamics between emissions, nuclear energy, renewable energy, and economic growth. Ecological Economics, 69(11), 2255-2260.
5. Azam, A., Rafiq, M., Shafique, M., Zhang, H., & Yuan, J. (2021). Analyzing the effect of natural gas, nuclear energy and renewable energy on GDP and carbon emissions: A multi-variate panel data analysis. Energy, 219, 119592.
6. Bang, G. (2010). Energy security and climate change concerns: Triggers for energy policy change in the United States? Energy policy, 38(4), 1645-1653.
7. Beane, S. R., Bedaque, P. F., Savage, M. J., & van Kolck, U. (2002). Towards a perturbative theory of nuclear forces. Nuclear Physics A, 700(1-2), 377-402.
8. Bethe, H. A. (1979). The fusion hybrid. Physics Today, 32(5), 44-51.
9. Bloch, I., Hunter, A., Appriou, A., Ayoun, A., Benferhat, S., Besnard, P., ... & Sossai, C. (2001). Fusion: General concepts and characteristics. International journal of intelligent systems, 16(10), 1107-1134.
10. Bock, R., Chu, Y. T., Dakowski, M., Gobbi, A., Grosse, E., Olmi, A., ... & Morenzoni, E. (1982). Dynamics of the fusion process. Nuclear Physics A, 388(2), 334-380.
11. Bowman, C. D., Arthur, E. D., Lisowski, P. W., Lawrence, G. P., Jensen, R. J., Anderson, J. L., ... & Wilson, W. B. (1992). Nuclear energy generation and waste transmutation using an accelerator-driven intense thermal neutron source. Nuclear Instruments and Methods in Physics Research Section A: Accelerators, Spectrometers, Detectors and Associated Equipment, 320(1-2), 336-367.

12. Çakar, N. D., Erdoğan, S., Gedikli, A., & Öncü, M. A. (2022). Nuclear energy consumption, nuclear fusion reactors and environmental quality: The case of G7 countries. Nuclear Engineering and Technology, 54(4), 1301-1311.
13. Chadwick, M. B., Herman, M., Obložinský, P., Dunn, M. E., Danon, Y., Kahler, A. C., ... & Young, P. G. (2011). ENDF/B-VII. 1 nuclear data for science and technology: cross sections, covariances, fission product yields and decay data. Nuclear data sheets, 112(12), 2887-2996.
14. Chadwick, M. B., Obložinský, P., Herman, M., Greene, N. M., McKnight, R. D., Smith, D. L., ... & Van der Marck, S. C. (2006). ENDF/B-VII. 0: next generation evaluated nuclear data library for nuclear science and technology. Nuclear data sheets, 107(12), 2931-3060.
15. Corradini, M. L., Kim, B. J., & Oh, M. D. (1988). Vapor explosions in light water reactors: A review of theory and modeling. Progress in Nuclear Energy, 22(1), 1-117.
16. Davydov, R., Antonov, V., Makeev, S., Batov, Y., Dudkin, V., & Myazin, N. (2019). New high-speed system for controlling the parameters of a nuclear reactor in a nuclear power plant. In E3S Web of Conferences. EDP Sciences. (Vol. 140, pp. 2001-2003).
17. Dobran, F. (2012). Fusion energy conversion in magnetically confined plasma reactors. Progress in Nuclear Energy, 60, 89-116.
18. Durmayaz, A., & Yavuz, H. (2001). Exergy analysis of a pressurized-water reactor nuclear-power plant. Applied Energy, 69(1), 39-57.
19. Elsheikh, B. M. (2013). Safety assessment of molten salt reactors in comparison with light water reactors. journal of radiation research and applied sciences, 6(2), 63-70.
20. Entler, S., Horacek, J., Dlouhy, T., & Dostal, V. (2018). Approximation of the economy of fusion energy. Energy, 152, 489-497.
21. Freedman, L. (1981). Britain: the first ex-nuclear power? International Security, 6(2), 80-104.
22. Freudenburg, W. R., & Baxter, R. K. (1985). Nuclear reactions: public attitudes and policies toward nuclear power. Review of Policy Research, 5(1), 96-110.
23. Hayashi, M., & Hughes, L. (2013). The Fukushima nuclear accident and its effect on global energy security. Energy policy, 59, 102-111.
24. Horlick-Jones, T., Prades, A., & Espluga, J. (2012). Investigating the degree of "stigma" associated with nuclear energy technologies: A cross-cultural examination of the case of fusion power. Public Understanding of

Science, 21(5), 514-533.

25. Johnson Jr, J. H., & Zeigler, D. J. (1988). Post-Chernobyl nuclear reactions in the USA. Environment and Planning C: Government and Policy, 6(3), 289-310.
26. Kambe, M., & Uotani, M. (1998). Design and development of fast breeder reactor passive reactivity control systems: LEM and LIM. Nuclear Technology, 122(2), 179-195.
27. Kang, H. G., Kim, M. C., Lee, S. J., Lee, H. J., Eom, H. S., Choi, J. G., & Jang, S. C. (2009). An overview of risk quantification issues for digitalized nuclear power plants using a static fault tree. Nuclear Engineering and Technology, 41(6), 849-858.
28. Kobayashi, K., Kawashima, K., Ohashi, M., Tohkura, A., Saito, M., & Fujii-e, Y. (1998). Applicability evaluation to a MOX fueled fast breeder reactor for a self-consistent nuclear energy system. Progress in Nuclear Energy, 32(3-4), 681-688.
29. Koning, A. J., & Rochman, D. (2008). Towards sustainable nuclear energy: Putting nuclear physics to work. Annals of Nuclear Energy, 35(11), 2024-2030.
30. Koomey, J., & Hultman, N. E. (2007). A reactor-level analysis of busbar costs for US nuclear plants, 1970–2005. Energy Policy, 35(11), 5630-5642.
31. Kunze, J. F., Mahar, J. M., Giraud, K. M., & Myers, C. W. (2010, January). Underground nuclear energy complexes: technical and economic advantages. In ASME International Mechanical Engineering Congress and Exposition (Vol. 44298, pp. 493-498).
32. Kuteev, B. V., & Goncharov, P. R. (2020). Fusion–fission hybrid systems: yesterday, today, and tomorrow. Fusion Science and Technology, 76(7), 836-847.
33. Lenzen, M. (2008). Life cycle energy and greenhouse gas emissions of nuclear energy: A review. Energy conversion and management, 49(8), 2178-2199.
34. Metcalfe, M. P., & Mills, R. W. (2015). Radiocarbon mass balance for a Magnox nuclear power station. Annals of Nuclear Energy, 75, 665-671.
35. Petrescu, F. I., Apicella, A., Petrescu, R. V., Kozaitis, S., Bucinell, R., Aversa, R., & Abu-Lebdeh, T. (2016). Environmental protection through nuclear energy. American Journal of Applied Sciences, 13(9), 941-946.
36. Plompen, A. J., Cabellos, O., De Saint Jean, C., Fleming, M., Algora, A., Angelone, M., ... & Žerovnik, G. (2020). The joint evaluated fission

and fusion nuclear data library, JEFF-3.3. The European Physical Journal A, 56(7), 1-108.

37. Rabbi, M. F., & Sabharwal, M. (2018). Dynamics of Nuclear Energy Policies in India: A Case Study on the Emergence of Nuclear Safety Regulatory Authority. Indian Journal of Public Administration, 64(4), 664-685.

38. Rahmanta, M. A., Harto, A. W., Agung, A., & Ridwan, M. K. (2023). Nuclear Power Plant to Support Indonesia's Net Zero Emissions: A Case Study of Small Modular Reactor Technology Selection Using Technology Readiness Level and Levelized Cost of Electricity Comparing Method. Energies, 16(9), 3752.

39. Rosa, E. A., & Freudenberg, W. R. (1993). The historical development of public reactions to nuclear power: Implications for nuclear waste policy. Public reactions to nuclear waste: Citizens' views of repository siting, 4(3), 32-63.

40. Saouma, V. E., & Hariri-Ardebili, M. A. (2014). A proposed aging management program for alkali silica reactions in a nuclear power plant. Nuclear Engineering and Design, 277, 248-264.

41. Srivastava, G. D., & Kulkarni, R. D. (2015, January). Design & development of microcontroller based programmable ramp generator for AC-DC converter for simulating decay power transient in experimental facility for nuclear power plants. In 2015 International Conference on Nascent Technologies in the Engineering Field (ICNTE). IEEE. (Vol. 2, pp. 1-6).

42. Standring, J. (1966). Calculation of the graphite weight-loss in civil magnox and advanced gas-cooled reactors. Journal of Nuclear Energy. Parts A/B. Reactor Science and Technology, 20(3), 201-217.

43. Stohl, A., Seibert, P., Wotawa, G., Arnold, D., Burkhart, J. F., Eckhardt, S., ... & Yasunari, T. J. (2012). Xenon-133 and caesium-137 releases into the atmosphere from the Fukushima Dai-ichi nuclear power plant: determination of the source term, atmospheric dispersion, and deposition. Atmospheric Chemistry and Physics, 12(5), 2313-2343.

44. Sunal, C. S., & Sunal, D. W. (1999). Nuclear reactions: Studying peaceful applications in the middle and secondary school. The Social Studies, 90(4), 164-170.

45. Teller, E., Ishikawa, M., Wood, L., Hyde, R., & Nuckolls, J. (1996). Completely automated nuclear reactors for long-term operation II: Toward a concept-level point-design of a high-temperature, gas-cooled central power station system, part II. In Proceedings of the International

Conference on Emerging Nuclear Energy Systems, ICENES (Vol. 96, pp. 123-127).

46. Tews, I., Davoudi, Z., Ekström, A., Holt, J. D., & Lynn, J. E. (2020). New ideas in constraining nuclear forces. Journal of Physics G: Nuclear and Particle Physics, 47(10), 103001.
47. Toschi, R. (1997). Nuclear fusion, an energy source. Fusion Engineering and Design, 36(1), 1-8.
48. Turconi, R., Boldrin, A., & Astrup, T. (2013). Life cycle assessment (LCA) of electricity generation technologies: Overview, comparability and limitations. Renewable and Sustainable Energy Reviews, 28, 555-565.
49. Was, G. S., Petti, D., Ukai, S., & Zinkle, S. (2019). Materials for future nuclear energy systems. Journal of Nuclear Materials, 527, 151837.
50. West, G. M., McArthur, S. D., & Towle, D. (2011). Knowledge-directed characterization of nuclear power plant reactor core parameters. Nuclear engineering and design, 241(9), 4013-4025.
51. Yi-can, W. U. (2001). A fusion neutron source driven sub-critical nuclear energy system: a way for early application of fusion technology. Plasma Science and Technology, 3(6), 1085.
52. Zhou, S., & Zhang, X. (2010). Nuclear energy development in China: a study of opportunities and challenges. Energy, 35(11), 4282-4288.
53. Zinkle, S. J., Ott, L. J., Ingersoll, D. T., Ellis, R. J., & Grossbeck, M. L. (2002, January). Overview of materials technologies for space nuclear power and propulsion. In AIP Conference Proceedings. American Institute of Physics. (Vol. 608, pp. 1063-1073).

CHAPTER

# 6 Energy Transformation: Electric Generators

## Contents

## INTRODUCTION

In physics, the concept of energy refers to the inherent ability of a system to bring about alterations, irrespective of any constraints imposed by entropy. The alteration of the overall energy of systems can solely be achieved by adding or removing energy, as energy is a conserved quantity per the first law of thermodynamics. As per the principles of special relativity, alterations in energy will correspondingly result in modifications in the system's mass. Furthermore, the aggregate mass of a system serves as an indicator of its energy (Heinemann et al., 2022).

The energy within a given system has the potential to transform, resulting in its relocation to an alternative state. Energy can be harnessed in various states to perform various physical tasks. Alternatively, energy can be utilized in natural processes and machines to fulfill societal needs by offering services such as heat, light, or motion. An illustrative instance involves converting the latent chemical energy into a gasoline and oxygen mixture within an internal combustion engine. This conversion process entails the generation of thermal energy, subsequently harnessed and converted into propulsive energy, specifically kinetic energy responsible for the locomotion of a vehicle. A photovoltaic cell can transform solar radiation into electrical energy, illuminating a light bulb or energizing a computer (Lu et al., 2003).

A transducer is a commonly used term for a device that facilitates energy conversion from one form to another. In a broad sense, it is generally observed that the majority of energy forms, excluding thermal energy, have the potential to be transformed into any other form of energy, theoretically achieving an efficiency of 100%. Efficiencies of this nature can manifest in practical scenarios, specifically when the conversion of chemical potential energy to kinetic energy, and vice versa, is achieved in isolated systems (Liu et al., 2011).

Conversion of various energy forms into heat can be highly efficient. However, achieving perfect efficiency is only feasible in isolated systems. If the universe's boundaries encompass all that exists, then the universe would be the sole isolated system. Our understanding and technological capabilities do not enable us to construct a self-contained system derived from a subset of the universe (Zhang et al., 2022).

Even within isolated systems, imperfect efficiency can arise when the energy distribution is already partially dispersed among numerous quantum states accessible to a group of particles. These particles can explore various momentum and position states within the phase space. Under these conditions, the concept of entropy, which refers to the equalization of energy distribution within a given system, dictates that subsequent states of the system must exhibit a comparable level of energy evenness at a minimum. In the context of the entire universe, it is impossible to consolidate energy into a smaller number of states once it has dispersed into them (Unsihuay et al., 2007).

## 6.1 HISTORY OF ENERGY TRANSFORMATION FROM THE EARLY UNIVERSE

Throughout history, the universe has undergone various energy transformations, commonly characterized by distinct energy manifestations that have persisted since the Big Bang. The energy above forms are subsequently discharged and transformed into more dynamic forms, such as kinetic or radiant energy, in the presence of an appropriate triggering mechanism. A phenomenon known as direct energy conversion occurs when hydrogen, which was generated in the early phases of the universe, aggregates into astronomical objects such as planets. The process above entails directly converting gravitational potential energy into thermal energy. The extensive gaseous atmospheres of celestial bodies such as Saturn, Uranus, Jupiter, and Neptune undergo persistent contraction, generating significant heat. This heat plays a dominant role in governing the meteorological phenomena observed on these planets, including the formation of atmospheric bands, wind patterns, and intense storms (Higaki et al., 2013).

Analogous phenomena that are familiar involve converting energy from the initial cosmic explosion. This includes nuclear decay, which involves liberating energy from heavy isotopes such as uranium and thorium, which act as the original "reservoir" of energy. Uranium's energy is employed for rapid release in nuclear fission bombs, whereas similar stored energies within atomic nuclei are autonomously emitted during diverse manifestations of radioactive decay. In this process, the thermal energy generated by the radioactive decay of atoms within the Earth's core is rapidly transformed into heat energy. The increased temperatures have the potential to induce the uplift of mountains through the processes of plate tectonics and orogenesis. The progressive increase in the height of the landmass consequently indicates the accumulation of thermal energy in the form of gravitational potential energy. The conversion of stored potential energy into kinetic energy occurs in landslides after initiating a triggering event (Kaiser, 2016). Earthquakes are commonly recognized as events that release stored elastic potential energy within rocks, which is a manifestation of mechanical potential energy derived from the underlying radioactive heat sources. Hence, drawing from existing knowledge, it is comprehended that commonplace phenomena such as landslides and earthquakes entail the dissipation of energy that has been amassed as potential energy within the Earth's gravitational field or as elastic strain, which is a form of mechanical potential energy, within rocks. Before this, the energy linked to these phenomena had been confined within sizable atoms, which arose from the gravitational potentials that converted energy during the star formation process, leading to the genesis of these atoms and the subsequent retention of energy within them (Brown et al., 2004).

Through a succession of comparable sequential procedures that initiate from the inception of the cosmos, the fusion of hydrogen nuclei within the Sun gives rise to an additional reservoir of potential energy that originated during the epoch of the Big Bang. According to theoretical knowledge, the phenomenon of space expansion took place during that specific timeframe. At the same time, the cooling process of the universe transpired at a pace that impeded the full fusion of hydrogen into elements possessing higher atomic weights. This suggests that hydrogen is a storage medium for potential energy that can be released via nuclear fusion. The fusion process discussed herein is initiated by the thermal and gravitational forces that arise from the collapse of hydrogen clouds during stellar formation (Yanagida & Yoshimura, 1988).

Consequently, a fraction of the energy produced from fusion reactions is transformed into solar radiation. Upon reaching the Earth's surface, sunlight possesses the capacity to be transformed into gravitational potential energy. This phenomenon can manifest through diverse mechanisms, including snow avalanches or the vaporization of water from oceans, followed by its subsequent condensation at elevated elevations. Once the stored energy is discharged at a hydroelectric dam, it can effectively generate electricity by driving turbines and generators. Solar radiation is crucial in influencing various weather phenomena on Earth.

A hurricane is a notable example of a meteorological phenomenon that is impacted by solar activity. This phenomenon results from the gradual release of thermal energy accumulated over a prolonged period in expansive and unstable areas of the Warm Ocean. As a result, the abrupt release of energy generates short-lived yet highly vigorous atmospheric turbulence.

During a series of transformation processes, the potential energy originally stored during the Big Bang is discharged through intermediate events. The potential for energy storage exists in various forms, which can transform into more dynamic manifestations of energy subsequent to a certain period. During these instances, a specific type of energy transforms into diverse forms, including thermal energy (Brandenberger et al., 1983).

## 6.2 ELECTRIC GENERATOR

The electricity generation process involves using an electric generator, which functions to convert mechanical energy into electrical energy. Motors convert electrical power into mechanical energy, while generators perform the opposite process. These two devices exhibit numerous similarities in their functioning. The generator induces the movement of electrons within the windings, facilitating their flow through the external electrical circuit. This phenomenon can be likened to a water pump, which generates a continuous flow of water without actually generating it. The potential origins of mechanical energy encompass a range of possibilities, including but not limited to reciprocating or turbine steam engines, the utilization of water through turbines or waterwheels, internal combustion engines, wind turbines, hand cranks, compressed air, and various other sources of mechanical energy (Macknick et al., 2012).

**Figure 6.1.** Illustration of a modern steam turbine generator (Source: Kara Fitts, Creative Commons License).

**Figure 6.2.** Illustration of an alternator (Source: By CZmarlin — Christopher Ziemnowicz, releases all rights but a photo credit would be appreciated if this image is used anywhere other than Wikipedia. Please leave a note at Wikipedia here. Thank you! - Own work, Public Domain, https://commons.wikimedia.o, Creative Commons License).

**Figure 6.3.** Image of a generator in Zwevegem, West Flanders, Belgium (Source: Susan Lescih, Creative Commons License).

## 6.3 HISTORICAL DEVELOPMENTS OF ELECTRIC GENERATORS

Before establishing the connection between electricity and magnetism, inventors devised electrostatic generators that operated on the principles of electrostatics. The voltages generated exhibited a high magnitude, whereas the corresponding currents displayed a low magnitude. The operational mechanism entails using mobile electrically powered belts, plates, and disks to facilitate charge transportation toward an electrode with high potential. The charge was generated by employing one of two mechanisms:

- Electrostatic induction is a phenomenon in which an electric charge is induced in an object without direct contact with a charged object, typically through the influence of an electric field.
- The triboelectric effect refers to the phenomenon in which two insulating materials, upon coming into contact and then separating, acquire an electric charge due to the redistribution of electrons between them (Baharoon et al., 2015).

**Did You Know?**

The concept of electromagnetic induction was discovered by Micheal Faraday in 1831.

### 6.3.1 Jedlik's Dynamo

In 1827, Anyos Jedlik, an inventor from Hungary, initiated a sequence of experiments centered around electromagnetic rotating devices that he called electromagnetic self-rotors. The first version of the single-pole electric starter, developed and finalized from 1852 to 1854, incorporated electromagnetic elements within the fixed and rotating components. The individual in question developed the notion of the dynamo a minimum of six years before Siemens and Wheatstone yet refrained from seeking a patent due to the belief that others had already made similar realizations. The fundamental idea revolves around using two opposing electromagnets to generate the magnetic field surrounding the rotor instead of employing permanent magnets. Additionally, the principle of self-excitation was also discovered (Halacsy, 1971).

### 6.3.2 Faraday's Disk

During the period spanning from 1831 to 1832, Michael Faraday made a significant breakthrough by uncovering the fundamental principle underlying the functionality of electromagnetic generators. The principle, subsequently referred to as Faraday's law, states that a varying magnetic flux induces a potential difference across the terminals of an electrical conductor.

In addition, Faraday developed the first electromagnetic generator, commonly referred to as the 'Faraday disk', which operated as a homo-polar generator. The apparatus comprised a copper disc capable of rotation strategically placed between the poles of a horseshoe magnet. A modest direct current (DC) voltage was produced. The lack of efficiency in this design can be ascribed to the presence of counter flows of current that cancel each other out in areas unaffected by the magnetic field. The current induction occurs in close proximity to the magnet, while in regions outside the magnetic field's reach, the current flows in the opposite direction. The existence of counterflow imposes limitations on the power output to the pickup wires and produces surplus heat within the copper disc. Following this, homo-polar generators effectively resolved this concern by utilizing an arrangement of magnets positioned around the circumference of the disc to maintain a continuous and unidirectional flow of electric current (Tanaka & Okuno, 2016).

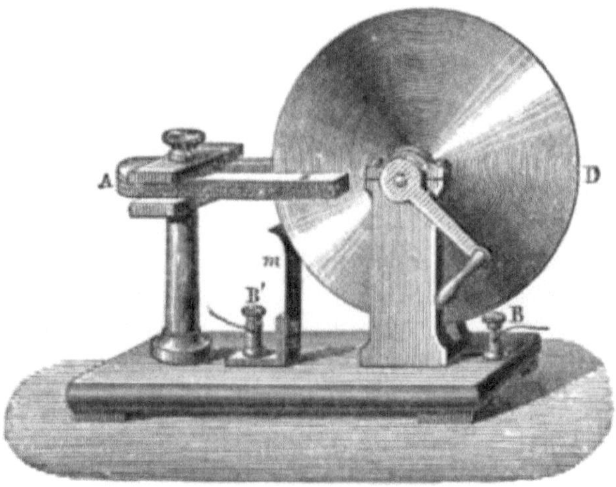


**Figure 6.4.** Illustration of Faraday Disk, the First Electric Generator. The horseshoe-shaped magnet (A) created a magnetic field through the disk (D). (Source: Lewis, Creative Commons License).

Another limitation was the lower output voltage from a single current pathway within the magnetic flux. The investigation conducted by researchers revealed that the utilization of multiple turns of wire in a coil resulted in increased and more viable voltages. The correlation between the output voltage and the number of turns in a generator facilitates the convenient manipulation of voltage levels by adjusting the number of turns. The integration of wire windings has emerged as an essential feature in all subsequent designs of generators (Crooks et al., 1978).

**Did You Know?**

The first practical electric generator was built by Hippolyte Pixxi in 1832. It was a hand-cracked device that produced alternating current (AC).

### 6.3.3 Dynamo

**Figure 6.5.** Illustration of the belt-driven dynamo (Source: By Hawkins Electrical Guide, Volume 7Chapter 66, Power Stations, Page 1979Copyright 1917 by Theo. Audel & Co.Printed in the United StatesScanned by Dale Mahalko, Gilman, WI, USA, Public Domain, https://commons.wikimedia.org/w/index.php?curid=4260515, Creative Commons License).

The utilization of dynamos for power generation has been discontinued primarily due to the challenges posed by the large dimensions and intricate nature of the commutator required for high-power applications. The dynamo above, which operates through a belt drive mechanism, exhibits a substantial capacity for generating electrical current. Specifically, it can produce 310 amperes at a voltage of 7 volts, resulting in a power output of 2,170 watts, all while rotating at 1400 RPM (Vaidyanathan et al., 2021).

The dynamo is historically recognized as the inaugural electrical generator capable of providing power for industrial purposes. The dynamo employs electromagnetic principles to transform mechanical rotation into a periodic direct electric current facilitated by a commutator. The initial dynamo was constructed by Hippolyte Pixii in the year 1832.

The dynamo, owing to a sequence of fortuitous discoveries, emerged as the precursor to numerous subsequent innovations, encompassing the DC electric motor, the AC alternator, the AC synchronous motor, and the rotary converter (Munir et al., 2020).

The prevalence of alternating current (AC) for power distribution and solid-state electronic AC to DC power conversion has significantly diminished the presence of large power generation dynamos in contemporary times. However, before the discovery of AC principles, the sole power generation and distribution method relied on significantly large direct-current dynamos. Currently, power generation dynamos are primarily regarded as objects of curiosity (Goto, 2014).

### 6.3.4 Other Rotating Electromagnetic Generators

The absence of a commutator in a dynamo results in the conversion of an alternator, which can be categorized as a synchronous singly-fed generator. In the context of providing electrical power to a grid, an alternator must maintain a consistent speed that is precisely synchronized with the electrical frequency of the power grid. A direct current (DC) generator can operate at different speeds within its mechanical constraints while consistently generating a waveform that exhibits direct current characteristics.

Implementing electronic control systems has the potential to improve the overall operational efficiency and effectiveness of generators significantly. Acknowledging that electronic control is exclusively integrated into doubly-fed or brushless wound-rotor doubly-fed generators is crucial. The power ratings of these generators are considerably lower than the power output of the controlled generator. Incorporating electronic control systems in these generators offers several cost, reliability, and efficiency advantages (Lauw & Meyer, 1982).

### 6.3.5 MHD Generator

The magnetohydrodynamic (MHD) generator is a piece of equipment that, by applying a magnetic field, makes it possible to directly extract electric power from the movement of hot gases, so doing away with the requirement for rotating electromagnetic equipment. Such generators were initially conceived to use the flame produced by a plasma MHD generator to heat the boilers of a steam power plant successfully. The AVCO Mk. 25, which underwent a redesign in 1965, was the first practical design to be introduced. It was named after its initials. The provision of major funding by the United States government in 1987 allowed for the construction a demonstration plant with a capacity of 25 megawatts (MW). The MHD plant U 25, located in the Soviet Union, was continuously connected to the Moscow power grid from 1972 until the late 1980s. This connection was maintained during the entirety of that time frame. The particular plant in question had a rating of 25 MW, making it the MHD plant with the highest global rating during that period. As of 2007, it has been noted that MHD generators, when operating in the topping cycle, demonstrate a lesser degree of efficiency compared to combined-cycle gas turbines (Kantrowitz et al., 1962).

**Figure 6.6.** Illustration of the rotor from the generator at Hoover Dam, United States (Source: Karas Fitts, Creative Commons License).

## 6.4 TERMINOLOGY

The generator or motor can be categorized into two primary components, elucidated from a mechanical or electrical perspective.

Mechanical:

- The rotor refers to an electrical machine's component responsible for rotation.
- The stator refers to the immobile component of an electrical machine.

Electrical:

- Armature: The component responsible for generating power in an electrical machine. The armature windings generate the electric current in a generator, alternator, or dynamo. The armature can be located on either the rotor or the stator.
- Field: The magnetic field aspect of an electrical machine. The magnetic field generation in a dynamo or alternator can be achieved through electromagnets or permanent magnets strategically positioned on either the rotor or the stator.

It is standard practice for alternating current (AC) generators to have the field winding on the rotor. This is because there is a large difference in the amount of power transferred between the field and armature circuits. On the other hand, the stator is where the winding for the armature is located. Slip rings are used, and just a small quantity of them is necessary to transfer field current to the rotating rotor. Direct current machines, also known as dynamos, require the presence of a commutator on the spinning shaft to convert the alternating current generated by the armature into direct current. This conversion takes place to make the machine more efficient. As a direct consequence, the armature winding of the machine is located on the rotor (Bosch & Groen, 2000).

## 6.5 EXCITATION

For an electric generator or motor to operate, the device must possess field coils instead of permanent magnets, necessitating an electric current within the field coils. Without a power supply to the field coils, the generator's rotor can rotate without generating any viable electrical energy, whereas the motor's rotor may remain stationary (Chan, 1995). On occasion, generators of a lesser size will display a trait known as self-excitation, characterized by the fact that the field coils will be powered by the electrical current created by the generator itself. It is possible to link the field coils to the armature winding in a series or a parallel configuration. When the generator is turned on for the first time, a magnetic field is generated due to the little residual magnetism inherent in the iron core. This magnetic field makes it easier for the generator to begin functioning. As a direct result, a very slight electric current is induced within the armature. A higher armature current is generated due to the electrical current being routed through the field coils, which boosts the strength of the magnetic field. The previously described "bootstrap" method continues until the core's magnetic field reaches saturation, ultimately resulting in the generator reaching a condition of consistent power generation (Tandon et al., 1984).

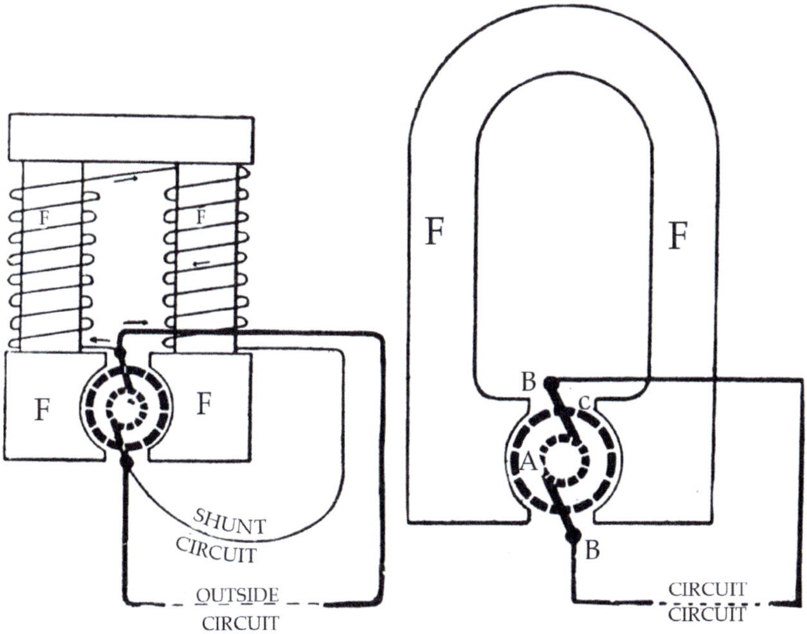


**Figure 6.7.** Illustration of a self exciter DC generator (Source: PD-US, https://en.wikipedia.org/w/index.php?curid=15183825, Creative Commons License).

**Remember**

The frequency of the electricity produced by generators can vary slightly depending on the speed of rotation and the load.

In many instances, significantly sized power station generators employ an additional, relatively smaller generator to stimulate the field coils of the larger generator. In the event of a severe and extensive power outage resulting in the isolation of power stations, it may become necessary for these stations to initiate a black start procedure. This procedure involves the activation of the largest generators within the stations to stimulate the magnetic fields, with the ultimate goal of reinstating power service to customers (Singh et al., 2006).

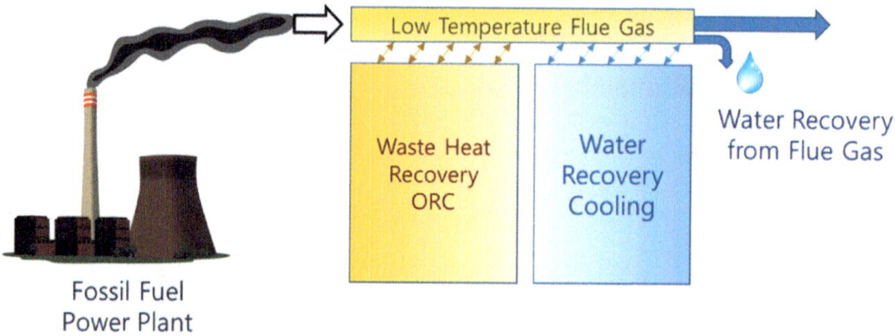


**Figure 6.8.** Representation of a power station (Source: Kim Y-M, Negash A, Shamsi SSM, Shin D-G, Cho G. Experimental Study of a Lab-Scale Organic Rankine Cycle System for Heat and Water Recovery from Flue Gas in Thermal Power Plants. Energies. 2021; 14(14):4328. https://doi.org/10.3390/en14144328, Creative Commons License).

Equivalent circuit of generator and load

G is generator

$V_G V_G$ Is its open circuit voltage

$R_G R_G$ is the internal resistance

$V_L V_L$ is on-load voltage

$R_L R_L$ is the load resistance

## 6.6 VEHICLE-MOUNTED GENERATORS

Historically, motor vehicles in the early years up until the 1960s predominantly employed direct current (DC) generators equipped with electromechanical regulators. Currently, alternators integrated with rectifier circuits have superseded the devices above due to their cost-effectiveness and reduced weight while providing comparable output. Automotive alternators supply electrical power to the various systems within a vehicle while also replenishing the battery's charge after the engine's ignition. The rated output of the electrical system within a vehicle is typically observed to fall within the range of 50 to 100 Amperes at a voltage of 12 Volts. This range is subject to variation based on the specific electrical load designed for the vehicle. Certain automobiles are currently equipped with electrically-driven steering assistance and air conditioning systems, resulting in a substantial electrical system burden. To provide adequate power to start a large diesel engine, it is common for large commercial vehicles to utilize a 24 V electrical system. Vehicle alternators do not employ permanent magnets and generally exhibit an efficiency range of 50-60% across a spectrum of speeds (Fu & Li, 2023). In motorcycle engineering, permanent magnet stators crafted from rare earth magnets are frequently employed in alternators due to their reduced size and weight advantages compared to alternative variants.

Bicycle lights are often powered by small generators that are commonly found. Typically, these devices are characterized by a current of 0.5 amperes and function as permanent-magnet alternators, providing an electrical power output ranging from 3 to 6 watts at 6 or 12 volts. Due to the reliance on the rider for power, there is a strong emphasis on efficiency in these systems. As a result, they often integrate rare-earth magnets and are meticulously designed and manufactured to ensure optimal performance. However, it should be noted that the highest achievable efficiency for these generators is approximately 80%. A higher prevalence of 60% can be attributed, at least partially, to rolling friction at the interface between the tire and the generator. This is primarily caused by inadequate alignment, relatively small generator size, bearing losses, and suboptimal design choices. Using permanent magnets results in a decline in efficiency at elevated velocities due to the inability to regulate the magnetic field intensity (Awal et al., 2015). Hub generators address several limitations as they are located within the bicycle hub itself, eliminating the need for a connection between the generator and the tire. In the past, acquiring these generators has proven costly and challenging. Prominent manufacturers of bicycle components, such as Shimano and SRAM, have recently ventured into this particular market. Nevertheless, it is anticipated that there will be notable advancements in the future due to the increasing popularity of cycling as a means of transportation and the development of LED technology, which enables the production of brighter lighting using the lower current output that these generators can supply (Wang et al., 2022).

## 6.7 ENGINE GENERATOR

An engine generator is a consolidated apparatus that integrates an electrical generator and an engine, creating a unified and self-contained unit. Typically, the propulsion systems employed consist of piston engines, although gas turbines may also be utilized. A wide array of versions is accessible, encompassing compact, portable petrol-powered sets and expansive turbine installations (Woo & Lee, 2014).

## 6.8 HUMAN POWERED ELECTRICAL GENERATORS

In specific contexts, a generator has the potential to be activated through human muscular force, as exemplified by its utilization in field radio station equipment (Paulides et al., 2009).

Human-powered direct current generators that are commercially available have been developed and have attracted the attention of individuals who engage in do-it-yourself (DIY) activities. Frequently propelled by pedal power, a modified bicycle trainer, or a foot pump, these generators can efficiently recharge batteries and, in specific cases, are outfitted with an integrated inverter. Based on empirical evidence, it has been observed that the average adult can generate a power output ranging from 125 to 200 watts through a pedal-powered generator. The purpose of incorporating a hand-crank mechanism in portable radio receivers is to reduce the necessity for frequent battery replacements (Yun et al., 2010).

## 6.9 LINEAR ELECTRIC GENERATOR

The most basic configuration of a linear electric generator involves the reciprocating motion of a magnet within a solenoid, which is essentially a coil of copper wire. Faraday's law of induction generates an alternating current within the wire loops whenever the magnet undergoes motion through them. The utilization of this particular generator is observed in the Faraday flashlight. Wave power schemes utilize larger linear electricity generators (Leijon et al., 2005).

## 6.10 TACHO GENERATOR

Tacho generators are commonly employed in powering tachometers, which measure the rotational speeds of electric motors, engines, and the associated machinery they drive. Generators produce an electrical potential that is approximately directly proportional to the rotational speed of the shaft. Generators can be engineered with meticulous construction and design to generate highly accurate voltages within specific ranges of rotational speeds (Ozgenel et al., 2017).

## SUMMARY

Electric generators are apparatuses that transform mechanical energy into electrical energy. Renewable energy sources play a pivotal and indispensable role in contemporary society by facilitating the provision of electricity across diverse domains. These sources supply power to residential and commercial establishments, propelling industrial equipment and generating electricity within power plants. Electric generators function based on the fundamental principle of electromagnetic induction. A conventional generator comprises multiple constituent parts, encompassing a rotor, a stator, a source of magnetic field, and a mechanism for transforming mechanical rotation into electrical output. Generators can be powered by a diverse range of energy sources, including but not limited to fossil fuels, nuclear energy, and renewable energy. Generators have profoundly impacted human society by facilitating the widespread accessibility of electricity, transforming various industries, and enhancing global quality of life.

## REFERENCES


1. Awal, M. R., Jusoh, M., Sakib, M. N., Hossain, F. S., Beson, M. R. C., & Aljunid, S. A. (2015). Design and implementation of vehicle mounted wind turbine. ARPN J Eng Appl Sci, 10, 8699-8860.
2. Baharoon, D. A., Rahman, H. A., Omar, W. Z. W., & Fadhl, S. O. (2015). Historical development of concentrating solar power technologies to generate clean electricity efficiently–A review. Renewable and Sustainable Energy Reviews, 41, 996-1027.
3. Bosch, J. L. H. R., & Groen, J. (2000). Sacral nerve neuromodulation in the treatment of patients with refractory motor urge incontinence: long-term results of a prospective longitudinal study. The Journal of Urology, 163(4), 1219-1222.
4. Brandenberger, R., Kahn, R., & Press, W. H. (1983). Cosmological perturbations in the early universe. Physical Review D, 28(8), 1809.
5. Brown, M. T., Odum, H. T., & Jorgensen, S. E. (2004). Energy hierarchy and transformity in the universe. Ecological Modelling, 178(1-2), 17-28.
6. Chan, T. F. (1995). Analysis of self-excited induction generators using an iterative method. IEEE transactions on energy conversion, 10(3), 502-507.
7. Crooks, M. J., Litvin, D. B., Matthews, P. W., Macaulay, R., & Shaw, J. (1978). One-piece Faraday generator: A paradoxical experiment from. Am. J. Phys, 46(7), 7.
8. Fu, Y., & Li, Y. (2023). Experimental Study on the Working Efficiency and Exergy Efficiency of the Vehicle-Mounted Thermoelectric Generator for Cold Chain Logistics Transportation Vehicle. Processes, 11(6), 1782.
9. Goto, H. (2014, March). Soliton electric generators based on conducting polymers. In IOP Conference Series: Materials Science and Engineering. IOP Publishing. (Vol. 54, pp. 120-121).
10. Halacsy, A. A. (1971). Anyos Jedlik, an inventor of the dynamo-electric principle. Electronics and Power, 17(9), 332-335.
11. Heinemann, G., Banzer, F., Dumitrescu, R., Hirschhausen, C. V., Neuhoff, M. E., & Nwadiaru, V. O. (2022). Transforming electricity access by replacing back-up generators with solar systems: Recent trends and evidence from Nigeria. Renewable and Sustainable Energy Reviews, 157, 111751.
12. Higaki, T., Nakayama, K., & Takahashi, F. (2013). Cosmological constraints on axionic dark radiation from axion-photon conversion in the early Universe. Journal of Cosmology and Astroparticle Physics, 2013(09), 030.

13. Kaiser, D. I. (2016). Nonminimal couplings in the early universe: multifield models of inflation and the latest observations. At the Frontier of Spacetime: Scalar-Tensor Theory, Bells Inequality, Machs Principle, Exotic Smoothness, 41-57.
14. Kantrowitz, A. R., Brogan, T. R., Rosa, R. J., & Louis, J. F. (1962). The magnetohydrodynamic power generator-basic principles, state of the art, and areas of application. IRE Transactions on Military Electronics, (1), 78-83.
15. Lauw, H. K., & Meyer, W. S. (1982). Universal machine modeling for the representation of rotating electric machinery in an electromagnetic transients program. IEEE Transactions on Power Apparatus and Systems, (6), 1342-1351.
16. Leijon, M., Bernhoff, H., Agren, O., Isberg, J., Sundberg, J., Berg, M., ... & Wolfbrandt, A. (2005). Multiphysics simulation of wave energy to electric energy conversion by permanent magnet linear generator. IEEE Transactions on Energy Conversion, 20(1), 219-224.
17. Liu, Z., Wen, F., Xue, Y., Xin, J., & Ledwich, G. (2011). Optimal siting and sizing of distributed generators considering plug-in electric vehicles. Dianli Xitong Zidonghua/Automation of Electric Power Systems, 35(18), 11-16.
18. Lu, F., Lee, H. P., & Lim, S. P. (2003). Modeling and analysis of micro piezoelectric power generators for micro-electromechanical-systems applications. Smart Materials and Structures, 13(1), 57.
19. Macknick, J., Newmark, R., Heath, G., & Hallett, K. C. (2012). Operational water consumption and withdrawal factors for electricity generating technologies: a review of existing literature. Environmental Research Letters, 7(4), 045802.
20. Mount, G. A. (1998). A critical review of ultralow-volume aerosols of insecticide applied with vehicle-mounted generators for adult mosquito control. Journal of the American Mosquito Control Association, 14(3), 305-334.
21. Munir, N., Khan, M., Wei, Z., Akgul, A., Amin, M., & Hussain, I. (2020). Circuit implementation of 3D chaotic self-exciting single-disk homopolar dynamo and its application in digital image confidentiality. Wireless Networks, 4, 1-18.
22. Ozgenel, M. C., Bal, G., & Uygun, D. (2017). Design and application of a novel high precision and low cost electronic tachogenerator for sensor-based brushless direct current motor drivers. Review of Scientific Instruments, 88(3), 251-255.
23. Paulides, J. J., Jansen, J. W., Encica, L. A. U. R. E. N. T. I. U., Lomonova,

E. A., & Smit, M. (2009, May). Human-powered small-scale generation system for a sustainable dance club. In 2009 IEEE International Electric Machines and Drives Conference. IEEE. (Vol. 1, pp. 439-444).

24. Singh, B., Murthy, S. S., & Gupta, S. (2006). Analysis and design of electronic load controller for self-excited induction generators. IEEE transactions on energy conversion, 21(1), 285-293.
25. Tanaka, M., & Okuno, Y. (2016). Plasma behavior in a high-temperature noble gas plasma disk-shaped MHD power generator. Electrical Engineering in Japan, 197(2), 33-40.
26. Tandon, A. K., Murthy, S. S., & Berg, G. J. (1984). Steady state analysis of capacitor self-excited induction generators. IEEE Transactions on Power Apparatus and Systems, (3), 612-618.
27. Unsihuay, C., Lima, J. M., & De Souza, A. Z. (2007, June). Modeling the integrated natural gas and electricity optimal power flow. In 2007 IEEE Power Engineering Society General Meeting. IEEE. (Vol. 2, pp. 1-7).
28. Vaidyanathan, S., Sambas, A., Abd-El-Atty, B., Abd El-Latif, A. A., Tlelo-Cuautle, E., Guillén-Fernández, O., ... & Ibrahim, M. A. H. (2021). A 5-D multi-stable hyperchaotic two-disk dynamo system with no equilibrium point: Circuit design, FPGA realization and applications to TRNGs and image encryption. Ieee Access, 9, 81352-81369.
29. Wang, X., Yuan, P., Li, Y., Chen, S., Shi, Z., & Xie, W. (2022, November). Performance research and optimization of vehicle-mounted claw-pole generator based on 3D electromagnetic field and circuit coupling modeling. In 2022 IEEE International Power Electronics and Application Conference and Exposition (PEAC). IEEE. (Vol. 2, pp. 886-891).
30. Woo, Y., & Lee, Y. J. (2014). Free piston engine generator: Technology review and an experimental evaluation with hydrogen fuel. International Journal of Automotive Technology, 15, 229-235.
31. Yanagida, T., & Yoshimura, M. (1988). Resonant axion-photon conversion in the early universe. Physics Letters B, 202(3), 301-306.
32. Yun, J., Patel, S. N., Reynolds, M. S., & Abowd, G. D. (2010). Design and performance of an optimal inertial power harvester for human-powered devices. IEEE Transactions on mobile computing, 10(5), 669-683.
33. Zhang, X., Chen, X., Qu, Y., Wu, Y., Wu, K., Deng, H., & Fu, Q. (2022). Fabricating high performance multi-functional hygroelectric generator through a biomimic approach. Nano Energy, 98, 107241.

CHAPTER

# 7 Energy Transformation: Geothermal Electricity and Fuel Cells

## Contents

# INTRODUCTION

Geothermal electricity is the electrical energy that is produced through the utilization of geothermal energy sources.

The range of estimates for the electricity generation capacity of geothermal energy spans from 35 to 2000 gigawatts. Geothermal power is widely regarded as a sustainable energy source due to its comparatively minimal heat extraction from the vast heat content of the Earth. The average emission intensity of electricity generated by already planted geothermal power plants is approximately 122 kg of carbon dioxide per megawatt-hour (Alirahmi et al., 2021).

A fuel cell facilitates the conversion of a primary fuel source into electrical energy through an electrochemical reaction. The process involves the production of electrical energy within a cell using chemical reactions between an oxidant and a fuel. The reactants are introduced into the cell, and the resulting reaction products are expelled from the cell (Rafikiran et al., 2022).

## 7.1 GEOTHERMAL ELECTRICITY

### 7.1.1 History and Development

The idea of geothermal power as a generating source emerged in the 20th century. On July 4, 1904, Prince Piero Ginori Conti conducted an experimental trial of the inaugural geothermal power generator in Larderello, Italy. It effectively illuminated four light bulbs. Subsequently, in 1911, the inaugural geothermal power plant designed for commercial purposes was constructed at that location. During the 1920s, generators were constructed in Beppu, Japan, and the Geysers, in California. However, it is worth noting that Italy was the sole industrial manufacturer of geothermal electricity worldwide until New Zealand established its own geothermal power plant in 1958 (Parri et al., 2021).

The year 1960 marked the commencement of operations by Pacific Gas and Electric of the inaugural geothermal electric power plant in the US, situated at The Geysers in California. The initial turbine demonstrated a lifespan exceeding three decades, yielding a net power output of 11 MW.

The initial demonstration of the binary cycle power plant occurred in 1967 within the Russian context, followed by its subsequent introduction to the United States in 1981. This technological advancement enables the utilization of significantly lower-temperature resources that were previously unattainable for recovery purposes. The year 2006 witnessed the commencement of operations of a binary cycle plant located in Chena Hot Springs, Alaska. This plant successfully generated electrical power by utilizing a fluid temperature as low as 57°C, setting a notable record in the field (Millstein et al., 2021).

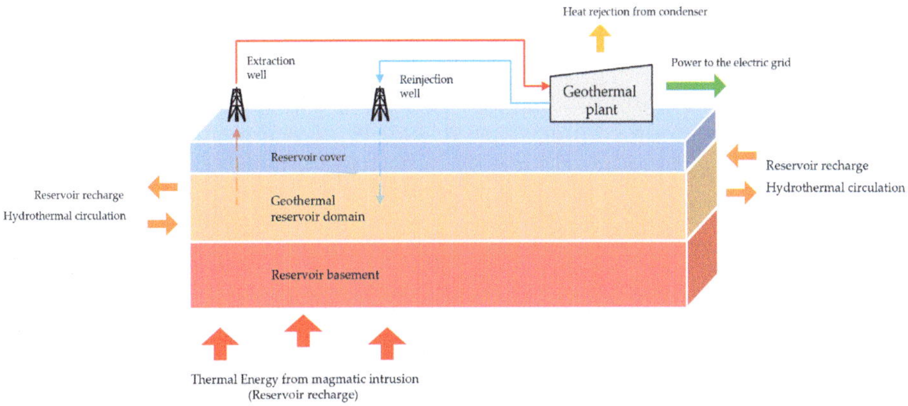


**Figure 7.1.** A scheme of a geothermal plant. (Source: Franco A, Vaccaro M. Sustainable Sizing of Geothermal Power Plants: Appropriate Potential Assessment Methods. Sustainability. 2020; 12(9):3844. https://doi.org/10.3390/su12093844, Creative Commons License).

Until recently, geothermal electric plants have been constructed solely in locations with abundant high-temperature geothermal resources near the Earth's surface.

The thermal efficiency of geothermal electric plants is relatively low, typically ranging from 10% to 23%. This is primarily attributed to the comparatively low temperature of geothermal fluids in comparison to the steam generated by conventional boilers. The efficiency of heat engines in harnessing useful energy during the electricity generation process is constrained by the low temperature in accordance with the principles of thermodynamics.

The impact of system efficiency on operational costs differs from that of coal, yet it remains a crucial consideration for the plant's overall feasibility. To achieve a net positive energy output, electricity generation necessitates utilizing high-temperature geothermal fields in conjunction with specialized heat cycles. Geothermal power exhibits a notable advantage over energy sources such as wind or solar due to its independence from variable energy inputs. This characteristic enables geothermal power to achieve a substantial capacity factor, as demonstrated by instances reaching up to 96%. In 2005, the global average stood at 73% (Quick et al., 2010).

### 7.1.2 Resources

The total amount of heat contained within the Earth is estimated to be approximately 1.031 x 10^31 joules. The heat is naturally transferred to the Earth's surface through conduction, occurring at a rate of 44.2 terawatts (TW). This heat is continuously replenished by the process of radioactive decay, which contributes an additional 30 TW of energy. The power rates mentioned exceed humanity's current energy consumption from primary sources by more than two-fold. However, the majority of this power is too diffuse, with an average of approximately 0.1 W/m2, making it unrecoverable (Barbier, 2002).

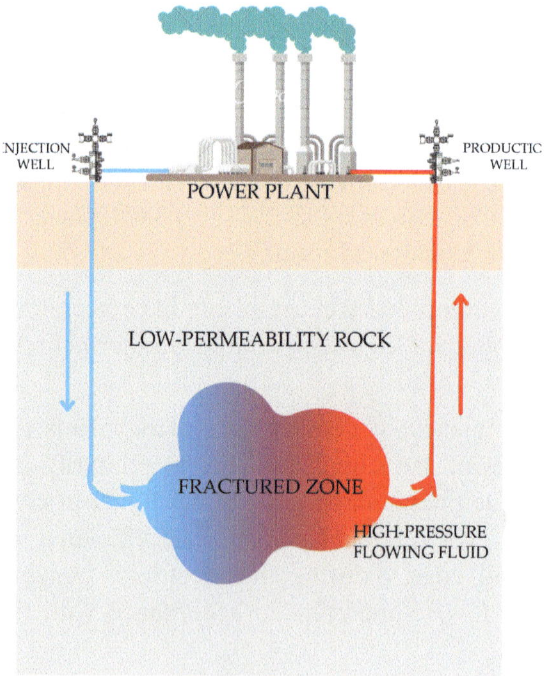


**Figure 7.2.** Illustration of an enhanced geothermal system (Source: Sowiżdżał A, Starczewska M, Papiernik B. Future Technology Mix—Enhanced Geothermal System (EGS) and Carbon Capture, Utilization, and Storage (CCUS)—An Overview of Selected Projects as an Example for Future Investments in Poland. Energies. 2022; 15(10):350, Creative Commons License).

The Earth's crust functions as a substantial insulating layer that necessitates the presence of fluid conduits, such as magma, water, or other substances, in order to facilitate the release of heat from beneath (Gutiérrez-Negrín, 2019).

The generation of electricity necessitates the utilization of high-temperature resources, which are exclusively attainable from subterranean depths. Heat transportation to the Earth's surface is facilitated by fluid circulation, which can occur through various mechanisms such as magma conduits, hot springs, hydrothermal circulation, oil wells, drilled water wells, or a combination of these processes. In certain instances, natural circulation occurs in regions characterized by a thin crust, whereby the transportation of magma through conduits facilitates the transfer of heat in proximity to the Earth's surface (Karimi & Mansouri, 2018). Additionally, the emergence of hot springs serves as a mechanism for the upward movement of this heat toward the surface. In the absence of a readily accessible hot spring, the alternative course of action involves drilling a well into a geothermal aquifer. In regions located distant from tectonic plate boundaries, it is generally observed that the geothermal gradient amounts to approximately 25-30°C per kilometer (km) of depth. Consequently, in order to facilitate the generation of electricity, it would be necessary for wells to extend to depths of several kilometers. The recoverable resources exhibit an enhanced combination of quantity and quality as the drilling depth increases and the proximity to tectonic plate boundaries decreases (Robson, 1974).

Introducing injected fluid can effectively enhance production in arid soil conditions or areas with insufficient water pressure. The developers initiate the site preparation process by drilling two holes into the designated area, subsequently employing either explosives or high-pressure water to fracture the rock formation between the holes. Subsequently, water or liquefied carbon dioxide is injected into a designated borehole and subsequently emerges as a gaseous form through another borehole. In Europe, this method is commonly referred to as hot, dry rock geothermal energy, while in North America, it is known as enhanced geothermal systems. This approach can potentially yield significantly greater benefits than conventional tapping of natural aquifers (Murphy & Niitsuma, 1999).

The electricity generating potential of geothermal energy exhibits a range of estimates, spanning from 35 to 2000 GW, contingent upon the magnitude of investments made. This analysis excludes non-electric heat sources such as co-generation, geothermal heat pumps, and other forms of direct utilization.

According to a report published in 2006 by the Massachusetts Institute of Technology (MIT), which examined the possibilities of enhanced geothermal systems, it was projected that a 15-year investment of 1 billion US dollars in research and development could result in the establishment of 100 GW of electrical generating capacity by 2050 solely within the United States. According to the MIT report, it has been estimated that a significant amount of over 200 zetta joules (ZJ) could be potentially extracted. Furthermore, this figure could potentially be increased to over 2,000 ZJ with technological advancements. Such a substantial energy reserve has the potential to meet the current global energy demands for multiple millennia (Bertani, 2009).

Currently, the depth of geothermal wells seldom exceeds 3 kilometers (approximately 2 miles). The upper bounds of geothermal resources are predicated on the assumption that wells can reach depths of up to 10 kilometers (equivalent to 6 miles). The ability to conduct drilling operations at this particular depth has become feasible within the petroleum industry, albeit at a significant cost. The Kola super deep borehole, which holds the distinction of being the deepest research well globally, reaches a depth of 12 kilometers (7 miles). This achievement has been replicated in the commercial sector by oil wells such as Exxon's Z-12 well located in the Chayvo field of Sakhalin. Drilling wells to depths exceeding 4 kilometers (equivalent to 2 miles) typically results in substantial drilling expenses amounting to tens of millions of dollars. The primary technological obstacles involve the efficient drilling of wide boreholes at a reduced expense, as well as the effective fragmentation of larger quantities of rock (Adams et al., 2021).

Geothermal power is widely acknowledged as a sustainable energy source due to its relatively minimal heat extraction in relation to the vast heat content of the Earth. However, it is imperative to exercise caution and closely monitor the extraction process to prevent any potential local depletion. While geothermal sites have the potential to supply heat for extended periods, it is important to note that individual wells may experience cooling or depletion of water resources. The production levels of the three oldest geothermal sites, namely Larderello, Wairakei, and the Geysers, have experienced a decline from their respective peak periods. The rate at which these plants extracted energy, relative to the rate at which it was replenished from deeper sources, as well as the potential depletion of the aquifers that provide them, remains uncertain. In the event of production reduction, coupled with the implementation of water reinjection, it is theoretically plausible for these wells to restore their productive capacity fully. Mitigation strategies have been previously implemented at certain locations. The Lardarello field in

Italy has provided evidence of the enduring viability of geothermal energy since 1913, followed by the Wairakei field in New Zealand since 1958 and The Geysers field in California since 1960 (Eberle et al., 2017).

## 7.1.3 Power Station Types

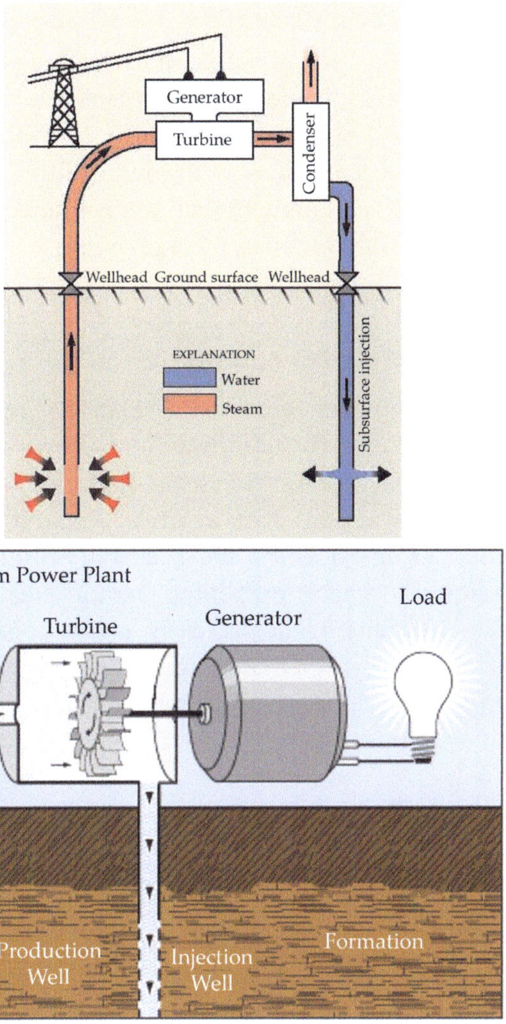


**Figure 7.3.** Illustration of the dry steam plant (Source: Abid K, Sharma A, Ahmed S, Srivastava S, Toledo Velazco A, Teodoriu C. A Review on Geothermal Energy and HPHT Packers for Geothermal Applications. Energies. 2022; 15(19):7357. https://doi.org/10.3390/en15197357, Creative Commons License).

#### *7.1.3.1 Dry Steam Power Plants*

Dry steam plants are considered the field's most basic and historically significant design. The turbines are operated by utilizing geothermal steam with a temperature of 150°C or higher (Rudiyanto et al., 2017).

#### *7.1.3.2 Flash Steam Power Plants*

Flash steam plants utilize the principle of extracting deep, high-pressure hot water and subsequently channeling it into tanks with lower pressure. This process enables steam generation through rapid depressurization, which is then employed to power turbines. A minimum fluid temperature of 180°C is typically required, although higher temperatures are often preferred. The plant under consideration is the prevailing variety in contemporary operational settings (Tomarov et al., 2019).

#### *7.1.3.3 Binary Cycle Power Plants*

Binary cycle power plants represent a relatively recent advancement in the field, as they can effectively utilize fluid temperatures as low as 57°C. The geothermal water, which possesses a moderate heat level, is circulated alongside a secondary fluid that exhibits a significantly lower boiling point than water. Consequently, the secondary fluid undergoes a phase change from liquid to vapor, propelling the turbines. The prevailing form of geothermal electricity plant currently under construction is the most frequently encountered. The Organic Rankine Cycle (ORC) and the Kalina Cycle are employed in various applications. The thermal efficiency typically exhibits a value of approximately 10% (Liu et al., 2014).

### 7.1.4 Worldwide Production

According to the International Geothermal Association (IGA) report, 10,715 megawatts (MW) of geothermal power is currently operational in 24 countries. This capacity is projected to generate approximately 67,246 gigawatt-hours (GWh) of electricity in 2010. This signifies a growth of 20% in the installed capacity of geothermal power since the year 2005. According to the International Geothermal Association (IGA), the geothermal energy capacity is projected to reach 18,500 MW by 2015. This growth is attributed to the significant number of projects currently being evaluated, particularly in regions that were previously believed to have limited exploitable geothermal resources (Biswal et al., 2015).

In the year 2010, the United States emerged as the global leader in geothermal electricity generation, boasting an impressive installed capacity of 3,086 MW across 77 power plants. Notably, the largest cluster of geothermal power plants worldwide is at The Geysers, a prominent geothermal field in California. The Philippines ranks second globally in terms of geothermal power production, following the United States. The country currently has an online capacity of 1,904 MW, with geothermal power contributing approximately 18% to the overall electricity generation in the nation (Hartman et al., 2011).

#### *7.1.4.1 Utility Grade Plants*

The Geysers, situated in California, United States, is home to the most extensive collection of geothermal power plants globally. As of 2004, it was observed that five specific countries, namely El Salvador, Kenya, the Philippines, Iceland, and Costa Rica, were able to produce a proportion of their electrical energy exceeding 15% through geothermal resources (Subedi et al., 2020).

The Naknek Electric Association (NEA) intends to commence drilling an exploratory well in the vicinity of King Salmon, located in the Southwest region of Alaska. The implementation of this technology has the potential to reduce electricity production costs by 71 percent. Furthermore, the proposed power capacity is estimated to be 25 megawatts (Hill et al., 2017).

Geothermal electricity is produced in the 24 countries enumerated in the tabular representation provided. In 2005, contracts were awarded to procure an additional 500 MW of electrical capacity within the United States. Furthermore, there were ongoing construction projects for power plants in 11 countries. Operational enhanced geothermal systems have been successfully implemented in France and Germany, with depths reaching several kilometers. Additionally, ongoing efforts are underway to develop or evaluate similar systems in at least four other countries (Guerin, 2017).

### 7.1.5 Environmental Impact

The fluids extracted from the Earth's depths contain a combination of various gases, including carbon dioxide (CO2), hydrogen sulfide (H2S), methane (CH4), and ammonia (NH3). These pollutants are known to contribute to global warming, acid rain, and the emission of noxious odors when released into the environment. Geothermal electric plants currently release an average of 122 kilograms of carbon dioxide (CO2) per megawatt-hour (MW·h) of electricity generated. This emission intensity is significantly lower compared to conventional fossil fuel power plants. Plants exposed to elevated levels of acidic compounds and volatile chemicals typically possess emission-control mechanisms to mitigate the release of pollutants. Theoretically, geothermal plants have the potential to sequester these gases by injecting them back into the Earth, thereby serving as a means of carbon capture and storage (Soltani et al., 2021).

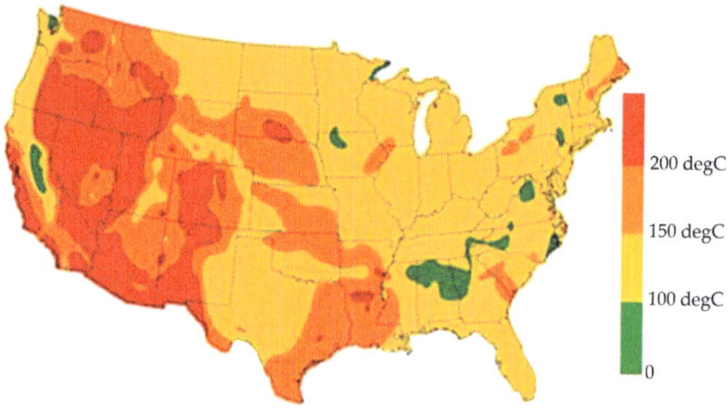


**Figure 7.4.** Geothermal resources n the US (Source: Abid K, Sharma A, Ahmed S, Srivastava S, Toledo Velazco A, Teodoriu C. A Review on Geothermal Energy and HPHT Packers for Geothermal Applications. Energies. 2022; 15(19):7357. https://doi.org/10.3390/en15197357, Creative Commons License).

Furthermore, alongside dissolved gases, hot water derived from geothermal sources can potentially contain minute quantities of hazardous substances, including mercury, arsenic, boron, antimony, and salt, which are solubilized within the water. These chemicals precipitate as the water temperature decreases, potentially leading to adverse environmental consequences upon their discharge. The contemporary method of reinjecting geothermal fluids into the Earth's subsurface to enhance production yields an additional advantage of mitigating environmental hazards (Paulillo et al., 2021).

The construction of plants has the potential to hurt the stability of the land. The Wairakei field in New Zealand has experienced subsidence. Enhanced geothermal systems have the potential to induce seismic activity through the process of hydraulic fracturing. The project in Basel, Switzerland, was halted due to over 10,000 seismic events within the initial 6-day period of water injection, with magnitudes reaching up to 3.4 on the Richter Scale. Geothermal energy exhibits limited demands for land and freshwater resources. Geothermal power plants require an area of 3.5 square kilometers to generate one gigawatt of electricity, specifically referring to the actual electrical production rather than the capacity. In contrast, coal facilities and wind farms necessitate areas of 32 and 12 square kilometers, respectively, for the same level of electrical output. The utilization of 20 liters of freshwater per MW·h is in stark contrast to the consumption of over 1000 liters per MW·h exhibited by nuclear, coal, or oil-based energy sources (Bošnjaković et al., 2019).

### 7.1.6 Economics

Geothermal power is characterized by its independence from fuel consumption, rendering it impervious to fluctuations in fuel costs. However, it is worth noting that geothermal power projects typically incur substantial capital expenses. Drilling costs constitute more than 50% of the overall expenses while exploring deep resources involves substantial risks. In Nevada, a conventional well doublet can facilitate the production of 4.5 megawatt (MW) of electrical power. The drilling expenses associated with such a well amount to approximately $10 million, while it is important to note that there is a 20% likelihood of encountering operational setbacks during the drilling process. The overall expenses for the construction of electrical plants and well drilling amount to approximately 2-5 million € per MW of electrical capacity. Concurrently, the levelized energy cost is estimated to range from 0.04-0.10 € per kW·h. Enhanced geothermal systems (EGS) typically exhibit higher values within these ranges, characterized by capital costs exceeding $4 million per MW and levelized costs surpassing $0.054 per kW·h in the year 2007 (Missimer et al., 2016).

Geothermal power exhibits significant scalability, as it can effectively cater to diverse energy demands. A considerable geothermal facility possesses the capacity to provide electricity for entire urban areas, while a smaller-scale power plant can adequately meet the energy requirements of a rural village.

Chevron Corporation is the foremost global private entity engaged in the production of geothermal electricity. The Geysers in California is widely recognized as the most highly developed geothermal field. In the year 2008, the sector above provided support to a total of 15 power plants, all of which were under the ownership of Calpine. These plants collectively possessed a generating capacity amounting to 725 megawatts (Clauser & Ewert, 2018).

## 7.2 FUEL CELL

Fuel cells exhibit a distinct characteristic compared to conventional electrochemical cell batteries, as they rely on the consumption of reactants obtained from an external source. This reliance necessitates the replenishment of these reactants, thereby operating within a thermodynamically open system. In contrast, batteries function as a means of storing electrical energy through chemical processes, thereby constituting a thermodynamically closed system.

There exists a wide range of potential combinations involving fuels and oxidants. A hydrogen fuel cell utilizes hydrogen as its primary fuel source, while oxygen, typically obtained from the surrounding air, serves as its oxidizing agent. Additional types of fuels encompass hydrocarbons and alcohols. Additional examples of oxidants include chlorine and chlorine dioxide (Carrette et al., 2000).

### 7.2.1 Design

Fuel cells are available in various types but operate using a similar fundamental mechanism. The structure consists of three distinct segments: the anode, the electrolyte, and the cathode, which are arranged in a sandwich-like configuration. Two chemical reactions take place at the interfaces of the three distinct segments. The two reactions' overall outcome is fuel consumption, water or carbon dioxide production, and an electric current, which can be harnessed to operate electrical devices, commonly known as the load (Owejan et al., 2009).

At the anode, a catalyst facilitates the oxidation of the fuel, typically hydrogen, resulting in the conversion of the fuel into a cation with a positive charge and an electron with a negative charge. The electrolyte is a substance that has been intentionally formulated to allow the passage of ions while impeding the flow of electrons. The liberated electrons traverse a conductor, thereby generating an electrical current. The ions traverse the electrolyte medium towards the cathode. Upon arrival at the cathode, the ions recombine with the electrons, resulting in a chemical reaction with a third substance, typically oxygen, leading to water or carbon dioxide formation (Zalc & Löffler, 2002).

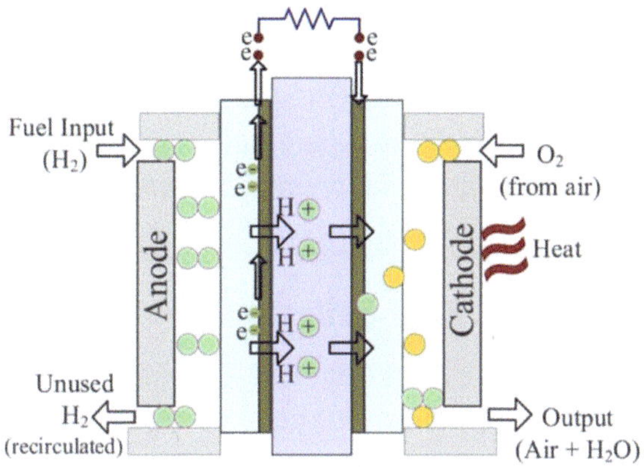


**Figure 7.5.** Block diagram of a proton exchange membrane fuel cell (PEMFC) (Source: Rezk H, Wilberforce T, Olabi AG, Ghoniem RM, Sayed ET, Ali Abdelkareem M. Optimal Parameter Identification of a PEM Fuel Cell Using Recent Optimization Algorithms. Energies. 2023; 16(14):5246. https://doi.org/10.3390/en16145246, Creative Commons License).

The primary design characteristics that hold the most importance in a fuel cell include:

- The electrolyte substance. The classification of fuel cells is typically determined by the electrolyte material employed.
- The type of fuel utilized. Hydrogen is the prevailing fuel source.
- The anode catalyst, which converts fuel into electrons and ions. The anode catalyst typically consists of finely divided platinum powder.
- The cathode catalyst facilitates the conversion of ions into waste chemicals such as water or carbon dioxide. Nickel is frequently employed as the cathode catalyst (Jung et al., 2014).

At full rated load, a conventional fuel cell typically generates a voltage ranging from 0.6 V to 0.7 V. Multiple factors contribute to the decrease in voltage as the current increases.

- Activation loss
- Ohmic loss (the voltage drop that occurs due to the resistance exhibited by the various components and interconnects within a cell).
- Mass transport loss (the phenomenon where reactants are depleted at catalyst sites under high loads, resulting in a rapid decline in voltage).

In order to achieve the desired energy output, the fuel cells can be interconnected in both series and parallel configurations. The series configuration results in an increased voltage, while the parallel configuration delivers a higher current. This particular configuration is commonly referred to as a fuel cell stack. One potential approach to enhance the current output of each cell is augmenting the cell surface area (Secanell et al., 2011).

### 7.2.2 Proton Exchange Fuel Cells

The conventional design of a proton exchange membrane fuel cell (PEMFC) involves the utilization of an archetypal hydrogen-oxygen system. This design employs a polymer membrane with proton-conducting properties as the electrolyte, effectively separating the anode and cathode compartments. During the early 1970s, the term "solid polymer electrolyte fuel cell" (SPEFC) was utilized to describe a specific type of fuel cell prior to the comprehensive understanding of the proton exchange mechanism. It is worth noting that the acronyms "PEM" and "PEM" are derived from the terms "polymer electrolyte membrane" and "proton exchange mechanism," respectively, despite their identical abbreviations (Li et al., 2009).

At the anode, hydrogen molecules diffuse towards the anode catalyst, subsequently dissociating into protons and electrons. The protons frequently engage in reactions with oxidants, leading to their transformation into what is commonly known as multi-facilitated proton membranes. The protons are transported across the membrane toward the cathode. At the same time, the electrons are compelled to traverse an external circuit in order to generate electrical power, as the membrane exhibits electrical insulating properties. In the context of the cathode catalyst, the process involves the interaction between oxygen molecules, electrons that have traversed the external circuit, and protons, resulting in water formation. It is important to note that water is the sole byproduct in this scenario, whether liquid or vapor (Devanathan, 2008).

In addition to the pure hydrogen variant, fuel cells can utilize hydrocarbon fuels such as diesel, methanol (specifically direct-methanol fuel cells and indirect methanol fuel cells), and chemical hydrides. The byproducts associated with the utilization of these fuel types encompass carbon dioxide and water (Sopian & Daud, 2006).

The PEMFC employs a bipolar plate as an electrode, featuring an in-milled gas channel structure. This particular electrode is fabricated using conductive plastics, which are further enhanced with carbon nanotubes to improve conductivity. Porous carbon papers are commonly used as a reactive layer, typically applied onto a polymer membrane (Peighambardoust et al., 2010). The materials utilized in fuel cells exhibit variations based on their respective types. In a conventional membrane electrode assembly (MEA), the bipolar plates utilized as electrodes are commonly constructed from metal, nickel, or carbon nanotubes. These plates are then coated with a catalyst, such as platinum, nano iron powders, or palladium, in order to enhance overall efficiency. The carbon paper serves as a physical barrier, effectively separating the electrolyte from the other components. The electrolyte has the potential to be either ceramic or a membrane (Wee, 2007).

### 7.2.3 Proton Exchange Membrane Fuel Cell Design Issues

- Costs. In the year 2002, the average cost of fuel cell systems amounted to US$1000 per kilowatt of electrical power generated. According to a report by the Department of Energy in 2009, it was stated that the cost of an automotive fuel cell system with a power output of 80 kW, when produced in large quantities (estimated at 500,000 units per year), amounts to $61 per kilowatt. The objective is to achieve a cost of $35 per kilowatt. In 2008, UTC Power offered stationary fuel cells with a capacity of 400 kW at a price of $1,000,000, covering the costs associated with their installation. The objective is to decrease expenses in order to effectively compete with existing market technologies, such as gasoline internal combustion engines. Numerous companies are currently engaged in endeavors to reduce costs through various means, one of which involves minimizing the quantity of platinum required in each cell. Ballard Power Systems conducted experiments using a carbon silk-enhanced catalyst, resulting in a noteworthy 30% decrease in platinum consumption (from 1 $mg/cm^2$ to 0.7 $mg/cm^2$) while maintaining performance levels unaltered. (Arvay et al., 2013).
- The production costs of the PEM (proton exchange membrane). The current market price for the Nafion membrane is $566 per square meter. In 2005, Ballard Power Systems made a public declaration regarding using Solupor, a patented porous polyethylene film developed by DSM, in their fuel cell technology.
- Water and air management (in PEMFCs). In this particular fuel cell variant, the membrane necessitates hydration, which entails

the evaporation of water at a rate that precisely matches its production. When water undergoes rapid evaporation, it leads to the desiccation of the membrane, resulting in an elevation of resistance across it. Over time, this elevated resistance can cause the membrane to develop cracks, thereby facilitating the occurrence of a gas "short circuit." This short circuit involves the direct combination of hydrogen and oxygen, generating heat that can inflict damage on the fuel cell. Suppose the rate of water evaporation is insufficiently rapid. In that case, an inundation of the electrodes may occur, impeding the reactants' access to the catalyst and consequently halting the progression of the reaction. Researchers are currently developing various methods for cellular water management, such as the utilization of electroosmotic pumps that specifically target flow control. Similar to a combustion engine, maintaining a consistent ratio between the reactant and oxygen is imperative for ensuring optimal efficiency in the operation of a fuel cell (Iranzo et al., 2020).

- Temperature management. In order to mitigate thermal loading and prevent cell destruction, it is imperative to maintain a consistent temperature throughout the entirety of the cell. The $2H_2 + O_2 \rightarrow 2H_2O$ reaction poses a significant challenge due to its highly exothermic nature, generating a substantial amount of heat within the fuel cell (Kurnia et al., 2021).
- Durability, service life, and special requirements for some types of cells. In stationary fuel cell applications, it is generally necessary to ensure a minimum of 40,000 hours of dependable operation within a temperature range of -35 °C to 40 °C (-31 °F to 104 °F). On the other hand, automotive fuel cells must be designed to withstand extreme temperatures and maintain a lifespan of 5,000 hours, which is equivalent to covering a distance of 150,000 miles. The present operational duration is 7,300 hours when subjected to cyclical circumstances. In order to meet the necessary requirements, automotive engines are expected to exhibit reliable starting capabilities even in extremely cold temperatures, specifically at -30°C (-22°F). Additionally, these engines must possess a high power-to-volume ratio, typically around 2.5 kilowatts per liter.
- Limited carbon monoxide tolerance of the cathode (Deng et al., 2022).

## 7.2.4 High-Temperature Fuel Cells

### *7.2.4.1 SOFC*

The solid oxide fuel cell (SOFC) offers significant advantages due to its potential for utilizing diverse fuel sources. In contrast to most fuel cells that exclusively rely on hydrogen as a fuel source, solid oxide fuel cells (SOFCs) can operate using a broader range of fuels, including hydrogen, butane, methanol, and various petroleum-based products. Each type of fuel possesses distinct chemical properties.

In the context of solid oxide fuel cells (SOFCs) utilizing methanol as a fuel source, the anode side of the cell is responsible for facilitating the catalytic breakdown of methanol and water into carbon dioxide, hydrogen ions, and free electrons. The hydrogen ions encounter oxide ions generated on the cathode and are transported through the electrolyte to the anode. At the anode, these ions undergo a reaction resulting in water formation. The electrical circuit is completed by a load that is externally connected between the anode and cathode. The following are the chemical equations representing the reaction (Mehr et al., 2021):

Anode Reaction: $C\,H_3\,H_3OH + H_2H_2O + 3O{=} \rightarrow CO_2O_2 + 3H_2H_2O + 6\underline{\underline{ee}}$

Cathode Reaction: $3/2\ O_2\ O_2 + 6e- \rightarrow 3O^{-}O^{-}$

Overall Reaction: $C\,H_3\,H_3OH + 3/2\ O_2O_2 \rightarrow CO_2O_2 + 2H_2H_2O +$ electrical energy

At the anode, solid oxide fuel cells (SOFCs) can employ nickel or other catalysts to facilitate the dissociation of methanol, resulting in hydrogen ions and carbon monoxide. The electrolyte employed in this application is a solid substance known as yttria-stabilized zirconia (YSZ). YSZ, similar to other fuel cell electrolytes, exhibits conductivity towards specific ions, specifically the oxide ion (O=), enabling its movement from the cathode to the anode. However, it lacks conductivity toward electrons. Yttria-stabilized zirconia (YSZ) is a highly resilient solid material with numerous benefits in large-scale industrial systems. While YSZ exhibits favorable ion conductivity, its functionality is limited to elevated temperatures. The accepted operating temperature is approximately 950 degrees Celsius. The elevated temperature at which the fuel cell operates facilitates the efficient dissociation of methane and oxygen molecules into ions. One significant drawback associated with the Solid Oxide Fuel Cell (SOFC) is the imposition of substantial limitations on selecting interconnection materials due to its elevated operating temperature. An additional drawback associated with operating the cell at elevated temperatures is the potential occurrence of undesired reactions within the fuel cell. The accumulation of carbon dust, specifically graphite, on the anode, is a prevalent occurrence that hinders the fuel's ability to reach the catalyst. Extensive research efforts are presently underway to explore alternative materials to yttria-stabilized zirconia (YSZ) that exhibit enhanced ion conductivity at reduced operating temperatures (Panopoulos et al., 2006).

#### *7.2.4.2 MCFC*

Molten carbonate fuel cells (MCFCs) function in a comparable manner, with the exception that the electrolyte is composed of liquid (molten) carbonate, which serves as a negative ion and an oxidizing agent. The electrolyte undergoes carbonate depletion during the oxidation reaction, necessitating the need for carbonate replenishment via certain mechanisms. This process is commonly executed by recirculating carbon dioxide from the oxidation byproducts. This recirculated carbon dioxide subsequently undergoes a reaction with the incoming air within the cathode, resulting in carbonate formation (Song, 2002).

In contrast to proton exchange fuel cells, the catalysts utilized in solid oxide fuel cells (SOFCs) and molten carbonate fuel cells (MCFCs) are not susceptible to carbon monoxide poisoning, primarily owing to their significantly elevated operational temperatures. Direct utilization of carbon monoxide is feasible due to the occurrence of the oxidation reaction at the anode. Moreover, the steam generated through the oxidation process can facilitate the displacement of carbon monoxide and the steam reformation of hydrocarbon fuels within the anode. These reactions can employ identical catalysts utilized in electrochemical reactions, thereby preventing the necessity for an external fuel reformer.

The utilization of MCFC (Molten Carbonate Fuel Cells) has the potential to effectively mitigate the emission of $CO_2$ from both coal-fired power plants and gas-turbine power plants (Ye et al., 2018).

### 7.2.5 Fuel Cell Efficiency

The efficiency of a fuel cell is contingent upon the magnitude of power extracted from it. Increasing power consumption results in a corresponding increase in current draw, thereby leading to elevated losses within the fuel cell. In general, it can be observed that there exists an inverse relationship between the amount of power (current) consumed and the level of efficiency achieved. The majority of losses in a cell are evident through a decrease in voltage, thus indicating that the efficiency of a cell is nearly directly proportional to its voltage. Hence, it is customary to present polarization curves depicting the relationship between voltage and current when discussing fuel cells. A cell operating at a voltage of 0.7 V exhibits an efficiency of approximately 50%, indicating that half of the hydrogen's energy content is converted into electrical energy. In contrast, the remaining half is dissipated as heat. The potential loss of fuel due to an incomplete reaction within the fuel cell system is contingent upon the specific design of the system (Taner, 2021).

In the case of a hydrogen cell functioning under standard conditions and without any reactant leakage, the efficiency can be determined by dividing the cell voltage by 1.48 V, considering the enthalpy or heating value associated with the reaction. The second law efficiency of a given cell is determined by dividing its cell voltage by 1.23 V. The cell's voltage is subject to variation based on factors such as the type of fuel utilized, the quality of the fuel, and the temperature conditions. The numerical disparity between these quantities signifies the distinction between the reaction's enthalpy and Gibbs free energy. The distinction above consistently manifests itself in the form of thermal energy in conjunction with any inefficiencies encountered during the electrical conversion process.

Fuel cells do not function based on a thermal cycle. Therefore, unlike combustion engines, they are not subject to the same thermodynamic constraints imposed by the Carnot cycle efficiency. Occasionally, there is a misrepresentation suggesting that fuel cells are not subject to the principles of thermodynamics. This misconception arises from the common association of thermodynamics with combustion processes, specifically the enthalpy of formation. The principles of thermodynamics are applicable to chemical processes, such as fuel cells, wherein the Gibbs free energy plays a significant role. It is worth noting that the maximum theoretical efficiency of these chemical processes, specifically the hydrogen/oxygen reaction in fuel cells, surpasses the thermal efficiency of the Otto cycle. For instance, at a temperature of 298K, the hydrogen/oxygen reaction exhibits an efficiency of 83%. In contrast, the Otto cycle, with a compression ratio of 10 and a specific heat ratio of 1.4, achieves a thermal efficiency of 60%. Utilizing thermodynamic limits as a predictive tool for practically attainable efficiencies is not considered reliable (Feroldi et al., 2009).

Moreover, in the pursuit of propulsion, it is necessary to convert the electrical output of the fuel cell into mechanical power, which incurs an additional decrease in efficiency. Regarding the exemption claim, it is accurate to assert that the constraints imposed by the second law of thermodynamics on the functioning of fuel cells are considerably less stringent compared to the limitations imposed on conventional energy conversion systems. As a result, fuel cells exhibit notable efficacy in the conversion of chemical energy into electrical energy, particularly when operating at low power density and utilizing pure hydrogen and oxygen as reactants. It is important to note that fuel cells, particularly those operating at high temperatures, have the potential to serve as a heat source within conventional heat engine systems, such as gas turbine systems. In this particular scenario, it is anticipated that the level of efficiency will be exceptionally high, surpassing 70% (Barbir & Gomez, 1997).

## 7.2.6 Fuel Cell Applications

### *7.2.6.1 Power*

Fuel cells are highly advantageous as sources of power in distant areas, including spacecraft, remote weather stations, expansive parks, rural regions, and specific military contexts. A hydrogen-powered fuel cell system has the potential to possess a compact and lightweight design while also being devoid of significant mechanical components. Fuel cells can attain reliability of up to 99.9999% under optimal circumstances due to their lack of moving components and absence of combustion. This corresponds to approximately one minute of downtime within a span of two years.

Electrolyzer systems, by their reliance on external storage units rather than internal fuel storage, exhibit potential for effective implementation in large-scale energy storage. This applicability is particularly evident in rural areas, which serve as a notable illustration. In this particular application, it is necessary to employ batteries of a significantly larger size to fulfill the storage requirements. Conversely, fuel cells necessitate only a larger storage unit, which is typically more cost-effective compared to an electrochemical device.

A pilot program is currently being implemented on Stuart Island in Washington State. The Stuart Island Energy Initiative has successfully implemented a comprehensive, self-contained system. Solar panels are utilized to generate electricity that is then used to power an electrolyzer, a device responsible for hydrogen production. The hydrogen is contained within a 500-gallon tank operating at a pressure of 200 pounds per square inch (PSI). This stored hydrogen is utilized by a ReliOn fuel cell system, which serves as a reliable source of electrical backup for the off-grid residence (Wilberforce et al., 2016).

### *7.2.6.2 Land vehicles*

The Hydrogen Fuel Initiative (HFI) was introduced by President George Bush in 2003 and subsequently enacted into law through the 2005 Energy Policy Act and the 2006 Advanced Energy Initiative. The primary objective of this initiative is to advance the research and development of hydrogen fuel cells and associated infrastructure technologies, with the ultimate aim of achieving the production of commercially viable fuel cell vehicles by the year 2020. As of 2008, the United States had made a financial contribution of 1 billion dollars towards this project.

In May 2009, the Obama Administration decided to cease funding for the advancement of fuel cell hydrogen vehicles, citing the potential for alternative vehicle technologies to achieve emissions reduction faster within a shorter timeframe. According to the US Secretary of Energy, the feasibility of hydrogen vehicles within the next 10 to 20 years is limited, as stated during a recent discussion. The Secretary also highlighted the complexities associated with establishing the necessary infrastructure for hydrogen fuel distribution. However, it is expected that the United States government will persist in providing financial support for research pertaining to stationary fuel cells. The decision to reduce funding for the Department of Energy's hydrogen and fuel cell program has been criticized by the National Hydrogen Association and the U.S. Fuel Cell Council. They argue that these proposed cuts pose a threat to the commercialization of a range of technologies that have demonstrated significant potential and are starting to gain momentum in the market (Khaligh & Li, 2010).

Motor vehicle manufacturers are currently engaged in extensive research and production of various prototype and production cars and buses that utilize fuel cell technology.

The GM 1966 Electro-van marked the initial endeavor of the automotive industry to develop a hydrogen fuel cell-powered vehicle. The Electro-van, with a weight exceeding twice that of a typical van, demonstrated a capacity to achieve speeds of up to 70 mph for 30 seconds. The 2001 Chrysler Natrium vehicle employed an independent hydrogen processor integrated into its onboard system. The hydrogen production for the fuel cell is achieved through the chemical reaction between sodium borohydride fuel and Borax, as stated by Chrysler, who asserted that both substances are abundantly found in the natural resources of the United States. The utilization of hydrogen in the fuel cell enables the generation of electrical energy, resulting in a nearly noiseless operation and a driving range of 300 miles, all while maintaining the integrity of the passenger compartment. Chrysler also engaged in the development of vehicles that implemented a hydrogen separation process within the vehicle itself. The primary objective of this endeavor was to mitigate emissions without being dependent on a non-existent hydrogen infrastructure while simultaneously circumventing the need for substantial storage tanks.

In 2005, Intelligent Energy, a British company, successfully developed the ENV (Emission Neutral Vehicle), which marked the first operational hydrogen-powered motorcycle. The motorcycle possesses a fuel capacity that sustains its operation for a duration of four hours, enabling it to traverse a distance of 100 miles within an urban setting while maintaining a maximum velocity of 50 miles per hour. In the year 2004, Honda successfully engineered a fuel cell motorcycle that incorporated the innovative Honda FC Stack technology (Sürer & Arat, 2022).

The inaugural Revolve Eco-Rally, initiated by HRH Prince of Wales, showcased a fleet of fuel cell vehicles on British roadways in 2007. Esteemed individuals and public figures, including celebrities and dignitaries, embarked on a journey from Brighton to Trafalgar Square in London, marking the first-ever public demonstration of such vehicles in the United Kingdom. The inaugural 2008 Formula Zero Championship marked the commencement of a groundbreaking hydrogen race series, where university students from various countries showcased their expertise in designing and constructing fuel-cell-powered race vehicles. This global event occurred in Rotterdam, the Netherlands, on August 22, 2008. Additional races are scheduled to take place in the years 2009 and 2010. Following the inaugural race, the team known as Greenchoice Forze, hailing from the esteemed University of Delft in the Netherlands, emerged as the frontrunner in the competition. Several other teams are competing, namely Element One from Detroit, HerUCLAs from Los Angeles, EUPLAtecH2 from Spain, Imperial Racing Green from London, and Zero Emission Racing Team from Leuven.

The FCX Clarity, a hydrogen vehicle, was introduced by Honda in 2008. In addition, alternative instances of bicycles and motorcycles are equipped with a hydrogen fuel cell propulsion system. Several companies are currently engaged in research on hydrogen fuel cells and are also conducting practical trials of fuel cell buses. In January 2007, Daimler AG conducted a three-year trial involving thirty-six experimental units powered by fuel cells manufactured by Ballard Power Systems. This trial took place in eleven cities and was deemed successful.

In addition, fuel cell-powered buses are presently operational or under development. One notable example is the fleet of Thor buses equipped with UTC Power fuel cells, which are currently in use by the SunLine Transit Agency in California. The Fuel Cell Bus Club is an international collaborative initiative focused on conducting trials of fuel cell buses.

In Brazil, a hydrogen-powered bus was produced in Caxias do Sul. At the same time, the hydrogen fuel will be generated in São Bernardo do Campo through electrolysis, which involves the decomposition of water molecules. The initiative, named "Ônibus Brasileiro a Hidrogênio" (Brazilian Hydrogen Autobus), incorporates three supplementary buses (Katalenich & Jacobson, 2022).

#### *7.2.6.3 Airplanes*

In February 2008, a consortium of Boeing researchers and industry partners in Europe conducted experimental flight tests on a piloted aircraft that relied solely on a fuel cell and lightweight batteries for propulsion. The Fuel Cell Demonstrator Airplane aircraft utilized a hybrid system consisting of a Proton Exchange Membrane (PEM) fuel cell and a lithium-ion battery. This system was employed to provide power to an electric motor, which in turn was connected to a conventional propeller. The inaugural flight of an aircraft propelled solely by a fuel cell, utilizing propeller propulsion, took place in the year 2003. The fuel cell employed a distinctive FlatStackTM stack configuration, enabling seamless integration of the fuel cell with the aerodynamic surfaces of the aircraft (Pratt et al., 2013).

#### *7.2.6.4 Fueling Stations*

In April 2003, the inaugural hydrogen refueling station accessible to the general public was established in Reykjavík, Iceland. This particular station accommodates three buses manufactured by DaimlerChrysler, which are currently operational within the public transportation network of Reykjavík. The hydrogen required by the station is self-generated through an electrolyzing unit manufactured by Norsk Hydro. The station operates autonomously and does not require external refilling, as it solely relies on the input of electricity and water. The project also includes Royal Dutch Shell as a collaborating entity. The absence of a roof at the station is intentional, as it facilitates the release of any hydrogen that may have leaked into the atmosphere.

The California Hydrogen Highway is an initiative spearheaded by the Governor of California to establish a network of hydrogen refueling stations throughout the state. These refueling stations are utilized to replenish hydrogen-powered vehicles, including fuel cell vehicles and hydrogen combustion vehicles. As of July 2007, the state of California possessed a total of 179 fuel cell vehicles, with twenty-five operational fueling stations. Additionally, there were plans to construct ten additional fueling stations in California. Nevertheless, three hydrogen fueling stations have already undergone decommissioning (Ilinich et al., 2007).

The state of South Carolina is currently in the process of developing a hydrogen freeway. Currently, two hydrogen fueling stations are located exclusively in Aiken and Columbia, South Carolina. Anticipated expansions in South Carolina include the establishment of supplementary stations in various locations, including Charleston, Myrtle Beach, Greenville, and Florence. As per the South Carolina Hydrogen & Fuel Cell Alliance, the current capacity of the Columbia station stands at 120 kg per day. Additionally, there are plans in place to further enhance the station's capabilities by implementing on-site hydrogen production through electrolysis and reformation methods. The current capacity of the Aiken station is 80 kilograms. South Carolina has allocated substantial financial resources toward advancing hydrogen fuel cell research and developing corresponding infrastructure. The University of South Carolina, an esteemed member of the South Carolina Hydrogen & Fuel Cell Alliance, was granted a substantial sum of 12.5 million dollars by the United States Department of Energy to support its Future Fuels Program.

Japan has implemented a hydrogen highway within the framework of the Japan hydrogen fuel cell project. A total of twelve hydrogen fueling stations have been constructed across eleven cities in Japan. In addition to the countries above, Canada, Sweden, and Norway have also successfully implemented hydrogen highways (Hwang, 2013).

### 7.2.7 Fuel Cell Economics

The utilization of hydrogen as a fuel source for vehicles would constitute a pivotal aspect within the framework of a hydrogen-based economy. The efficiency of a fuel cell and electric motor combination is not subject to the same limitations imposed by the Carnot efficiency of an internal combustion engine.

Platinum catalysts are utilized in low-temperature fuel cell stacks, including proton exchange membrane fuel cells (PEMFCs), direct methanol fuel cells (DMFCs), and phosphoric acid fuel cells (PAFCs). The presence of impurities in low-temperature fuel cells leads to catalyst poisoning, decreasing activity and efficiency. Consequently, ensuring high hydrogen purity or employing catalysts with higher densities is necessary. Despite the presence of ample platinum resources to meet future demand, many predictions regarding the depletion of platinum reserves and subsequent price surges fail to consider the potential impact of catalyst loading reduction and recycling practices. A recent study at Brookhaven National Laboratory explored the potential of utilizing a gold-palladium coating as a substitute for platinum in fuel cells. This alternative coating has shown promise in mitigating the issue of poisoning, which could significantly enhance the longevity of fuel cells. An alternative approach could involve the utilization of iron and sulfur in lieu of platinum. This phenomenon can be achieved through an intermediary process facilitated by bacteria. The cost reduction of a fuel cell would be significant, as the price of platinum in a conventional fuel cell is approximately \$1500, whereas an equivalent amount of iron costs only around \$1.50. The concept is currently being developed through a collaborative effort between the John Innes Centre and the University of Milan-Bicocca.

The present objectives for transport proton exchange membrane (PEM) fuel cells involve achieving a platinum (Pt) loading of 0.2 g/kW, which represents a five-fold reduction compared to current loadings. Prominent original equipment manufacturers (OEMs) have expressed optimism regarding the feasibility of attaining this target. The conservation of supplies can be achieved by recycling fuel cell components, such as platinum. High-temperature fuel cells, such as molten carbonate (MCFCs) and solid oxide fuel cells (SOFCs), employ cost-effective catalysts such as nickel and nickel oxide rather than platinum. Additionally, these catalysts exhibit resistance to carbon monoxide-induced poisoning, thereby preventing the need for hydrogen of high purity in their operation. One viable approach involves directly utilizing readily available fuels, such as natural gas, which possess a well-established and extensive infrastructure. This prevents the need for an external conversion process to generate hydrogen and carbon monoxide (CO), followed by subsequent removal of CO (Wasajja et al., 2020).

## SUMMARY

Geothermal electricity is a type of sustainable energy sourced from the thermal energy contained within the Earth's crust. Using geothermal energy involves extracting electricity using diverse techniques, commonly by accessing underground hot water and steam reservoirs. Geothermal power plants are typically situated in regions that offer convenient proximity to geothermal reservoirs. These phenomena are frequently observed in close proximity to tectonic plate boundaries, volcanic regions, and geothermal areas. Geothermal energy exhibits the advantage of being naturally replenished and impervious to depletion through utilization, thereby establishing itself as a dependable and sustainable energy source over extended periods.

Fuel cells are electrochemical devices that efficiently convert the chemical energy stored in a fuel, such as hydrogen, into electrical energy through a direct electrochemical reaction without the need for combustion. These energy conversion systems are widely recognized for their high efficiency and low emissions. Fuel cells have the potential to attain notable levels of conversion efficiencies, particularly when employed within combined heat and power systems. Fuel cells can be utilized across a range of applications, encompassing both compact portable devices and expansive stationary power plants, as well as their integration into vehicles.

## REFERENCES

1. Adams, B., Ogland-Hand, J., Bielicki, J. M., Schädle, P., & Saar, M. (2021). Estimating the geothermal electricity generation potential of sedimentary basins using genGEO (the generalizable GEOthermal techno-economic simulator) (Vol. 2, pp. 69-70).
2. Alirahmi, S. M., Assareh, E., Chitsaz, A., Holagh, S. G., & Jalilinasrabady, S. (2021). Electrolyzer-fuel cell combination for grid peak load management in a geothermal power plant: Power to hydrogen and hydrogen to power conversion. International Journal of Hydrogen Energy, 46(50), 25650-25665.
3. Arvay, A., French, J., Wang, J. C., Peng, X. H., & Kannan, A. M. (2013). Nature inspired flow field designs for proton exchange membrane fuel cell. International Journal of Hydrogen Energy, 38(9), 3717-3726.
4. Barbier, E. (2002). Geothermal energy technology and current status: an overview. Renewable and Sustainable Energy Reviews, 6(1-2), 3-65.
5. Barbir, F., & Gomez, T. (1997). Efficiency and economics of proton exchange membrane (PEM) fuel cells. International Journal of Hydrogen Energy, 22(10-11), 1027-1037.
6. Bertani, R. (2009, May). Geothermal energy: an overview on resources and potential. In Proceedings of the International Conference on National Development of Geothermal Energy Use, Slovakia.4(1), 412-415.
7. Biswal, A., Tripathy, B. C., Sanjay, K., Subbaiah, T., & Minakshi, M. (2015). Electrolytic manganese dioxide (EMD): a perspective on worldwide production, reserves and its role in electrochemistry. RSC Advances, 5(72), 58255-58283.
8. Bošnjaković, M., Stojkov, M., & Jurjević, M. (2019). Environmental impact of geothermal power plants. Tehnički vjesnik, 26(5), 1515-1522.
9. Carrette, L., Friedrich, K. A., & Stimming, U. (2000). Fuel cells: principles, types, fuels, and applications. ChemPhysChem, 1(4), 162-193.
10. Clauser, C., & Ewert, M. (2018). The renewables cost challenge: Levelized cost of geothermal electric energy compared to other sources of primary energy–Review and case study. Renewable and Sustainable Energy Reviews, 82, 3683-3693.
11. Deng, X., Huang, C., Pei, X., Hu, B., & Zhou, W. (2022). Recent progresses and remaining issues on the ultrathin catalyst layer design strategy for high-performance proton exchange membrane fuel cell with further reduced Pt loadings: A review. International Journal of Hydrogen Energy, 47(3), 1529-1542.

12. Devanathan, R. (2008). Recent developments in proton exchange membranes for fuel cells. Energy & Environmental Science, 1(1), 101-119.
13. Eberle, A., Heath, G. A., Carpenter Petri, A. C., & Nicholson, S. R. (2017). Systematic review of life cycle greenhouse gas emissions from geothermal electricity. National Renewable Energy Laboratory. 7, 58727.
14. Feroldi, D., Serra, M., & Riera, J. (2009). Energy management strategies based on efficiency map for fuel cell hybrid vehicles. Journal of Power Sources, 190(2), 387-401.
15. Guerin, T. (2017). A case study identifying and mitigating the environmental and community impacts from construction of a utility-scale solar photovoltaic power plant in eastern Australia. Solar Energy, 146, 94-104.
16. Gutiérrez-Negrín, L. C. (2019, March). Current status of geothermal-electric production in Mexico. In IOP Conference Series: Earth and Environmental Science. IOP Publishing. (Vol. 249, pp. 12017).
17. Hartman, G. L., West, E. D., & Herman, T. K. (2011). Crops that feed the World 2. Soybean—worldwide production, use, and constraints caused by pathogens and pests. Food Security, 3, 5-17.
18. Hill, D. J., Tarasoff, C., Whitworth, G. E., Baron, J., Bradshaw, J. L., & Church, J. S. (2017). Utility of unmanned aerial vehicles for mapping invasive plant species: a case study on yellow flag iris (Iris pseudacorus L.). International Journal of Remote Sensing, 38(8-10), 2083-2105.
19. Hwang, J. J. (2013). Sustainability study of hydrogen pathways for fuel cell vehicle applications. Renewable and Sustainable Energy Reviews, 19, 220-229.
20. Ilinich, O., Ruettinger, W., Liu, X., & Farrauto, R. (2007). Cu–Al2O3–CuAl2O4 water–gas shift catalyst for hydrogen production in fuel cell applications: Mechanism of deactivation under start–stop operating conditions. Journal of Catalysis, 247(1), 112-118.
21. Iranzo, A., Arredondo, C. H., Kannan, A. M., & Rosa, F. (2020). Biomimetic flow fields for proton exchange membrane fuel cells: A review of design trends. Energy, 190, 116435.
22. Jung, N., Chung, D. Y., Ryu, J., Yoo, S. J., & Sung, Y. E. (2014). Pt-based nanoarchitecture and catalyst design for fuel cell applications. Nano Today, 9(4), 433-456.
23. Karimi, S., & Mansouri, S. (2018). A comparative profitability study of geothermal electricity production in developed and developing countries: Exergoeconomic analysis and optimization of different ORC

configurations. Renewable Energy, 115, 600-619.

24. Katalenich, S. M., & Jacobson, M. Z. (2022). Toward battery electric and hydrogen fuel cell military vehicles for land, air, and sea. Energy, 254, 124355.
25. Khaligh, A., & Li, Z. (2010). Battery, ultracapacitor, fuel cell, and hybrid energy storage systems for electric, hybrid electric, fuel cell, and plug-in hybrid electric vehicles: State of the art. IEEE transactions on Vehicular Technology, 59(6), 2806-2814.
26. Kurnia, J. C., Chaedir, B. A., Sasmito, A. P., & Shamim, T. (2021). Progress on open cathode proton exchange membrane fuel cell: Performance, designs, challenges and future directions. Applied Energy, 283, 116359.
27. Li, Q., Jensen, J. O., Savinell, R. F., & Bjerrum, N. J. (2009). High temperature proton exchange membranes based on polybenzimidazoles for fuel cells. Progress in polymer science, 34(5), 449-477.
28. Liu, X., Wang, X., & Zhang, C. (2014). Sensitivity analysis of system parameters on the performance of the Organic Rankine Cycle system for binary-cycle geothermal power plants. Applied thermal engineering, 71(1), 175-183.
29. Mehr, A. S., Lanzini, A., Santarelli, M., & Rosen, M. A. (2021). Polygeneration systems based on high temperature fuel cell (MCFC and SOFC) technology: System design, fuel types, modeling and analysis approaches. Energy, 228, 120613.
30. Millstein, D., Dobson, P., & Jeong, S. (2021). The potential to improve the value of US geothermal electricity generation through flexible operations. Journal of Energy Resources Technology, 143(1), 010905.
31. Missimer, T. M., Choon Ng, K., Thuw, K., & Wakil Shahzad, M. (2016). Geothermal electricity generation and desalination: an integrated process design to conserve latent heat with operational improvements. Desalination and Water Treatment, 57(48-49), 23110-23118.
32. Murphy, H., & Niitsuma, H. (1999). Strategies for compensating for higher costs of geothermal electricity with environmental benefits. Geothermics, 28(6), 693-711.
33. Owejan, J. P., Gagliardo, J. J., Sergi, J. M., Kandlikar, S. G., & Trabold, T. A. (2009). Water management studies in PEM fuel cells, Part I: Fuel cell design and in situ water distributions. International Journal of Hydrogen Energy, 34(8), 3436-3444.
34. Panopoulos, K. D., Fryda, L. E., Karl, J., Poulou, S., & Kakaras, E. (2006). High temperature solid oxide fuel cell integrated with novel allothermal biomass gasification: Part I: Modelling and feasibility study. Journal of

Power Sources, 159(1), 570-585.

35. Parri, R., Basile, P., Favaro, L., Mazzoni, T., & Orlando, S. (2021). The history of geothermal electric power plants on the Island of Ischia, Italy. Geothermics, 89, 101977.
36. Paulillo, A., Kim, A., Mutel, C., Striolo, A., Bauer, C., & Lettieri, P. (2021). Influential parameters for estimating the environmental impacts of geothermal power: A global sensitivity analysis study. Cleaner Environmental Systems, 3, 100054.
37. Peighambardoust, S. J., Rowshanzamir, S., & Amjadi, M. (2010). Review of the proton exchange membranes for fuel cell applications. International Journal of Hydrogen Energy, 35(17), 9349-9384.
38. Pratt, J. W., Klebanoff, L. E., Munoz-Ramos, K., Akhil, A. A., Curgus, D. B., & Schenkman, B. L. (2013). Proton exchange membrane fuel cells for electrical power generation on-board commercial airplanes. Applied Energy, 101, 776-796.
39. Quick, H., Michael, J., Huber, H., & Arslan, U. (2010). History of international geothermal power plants and geothermal projects in Germany. In Proceedings World Geothermal Congress (Vol. 3, pp. 25-29).
40. Rafikiran, S., Basha, C. H., Murali, M., & Fathima, F. (2022). Design and performance evaluation of solid oxide-based fuel cell stack for electric vehicle system with modified marine predator optimized fuzzy controller. Materials Today: Proceedings, 60, 1898-1904.
41. Robson, G. R. (1974). Geothermal electricity production. Science, 184(4134), 371-375.
42. Rudiyanto, B., Illah, I., Pambudi, N. A., Cheng, C. C., Adiprana, R., Imran, M., ... & Handogo, R. (2017). Preliminary analysis of dry-steam geothermal power plant by employing exergy assessment: Case study in Kamojang geothermal power plant, Indonesia. Case studies in thermal engineering, 10, 292-301.
43. Secanell, M., Wishart, J., & Dobson, P. (2011). Computational design and optimization of fuel cells and fuel cell systems: A review. Journal of Power Sources, 196(8), 3690-3704.
44. Soltani, M., Kashkooli, F. M., Souri, M., Rafiei, B., Jabarifar, M., Gharali, K., & Nathwani, J. S. (2021). Environmental, economic, and social impacts of geothermal energy systems. Renewable and Sustainable Energy Reviews, 140, 110750.
45. Song, C. (2002). Fuel processing for low-temperature and high-temperature fuel cells: Challenges, and opportunities for sustainable development in the 21st century. Catalysis Today, 77(1-2), 17-49.

46. Sopian, K., & Daud, W. R. W. (2006). Challenges and future developments in proton exchange membrane fuel cells. Renewable Energy, 31(5), 719-727.

47. Subedi, P., Gattoni, K., Liu, W., Lawrence, K. S., & Park, S. W. (2020). Current utility of plant growth-promoting rhizobacteria as biological control agents towards plant-parasitic nematodes. Plants, 9(9), 1167.

48. Sürer, M. G., & Arat, H. T. (2022). Advancements and current technologies on hydrogen fuel cell applications for marine vehicles. International Journal of Hydrogen Energy, 47(45), 19865-19875.

49. Taner, T. (2021). The novel and innovative design with using H2 fuel of PEM fuel cell: Efficiency of thermodynamic analyze. Fuel, 302, 121109.

50. Tomarov, G. V., Borzenko, V. I., & Shipkov, A. A. (2019). Optimization investigations of a combined binary-cycle geothermal power plant with two separation pressures and flashed steam superheating using a hydrogen–oxygen steam generator. Thermal Engineering, 66, 760-768.

51. Wasajja, H., Lindeboom, R. E., van Lier, J. B., & Aravind, P. V. (2020). Techno-economic review of biogas cleaning technologies for small scale off-grid solid oxide fuel cell applications. Fuel Processing Technology, 197, 106215.

52. Wee, J. H. (2007). Applications of proton exchange membrane fuel cell systems. Renewable and Sustainable Energy Reviews, 11(8), 1720-1738.

53. Wilberforce, T., Alaswad, A., Palumbo, A., Dassisti, M., & Olabi, A. G. (2016). Advances in stationary and portable fuel cell applications. International Journal of Hydrogen Energy, 41(37), 16509-16522.

54. Ye, Z., Zhang, X., Li, W., Su, G., & Chen, J. (2018). Optimum operation states and parametric selection criteria of a high temperature fuel cell-thermoradiative cell system. Energy Conversion and Management, 173, 470-475.

55. Zalc, J. M., & Löffler, D. G. (2002). Fuel processing for PEM fuel cells: transport and kinetic issues of system design. Journal of Power Sources, 111(1), 58-64.

CHAPTER

# 8 Future Trends in Energy

## Contents

## INTRODUCTION

This chapter will examine the anticipated structure of the energy economy in forthcoming years. This statement does not constitute a prediction due to the inherent difficulty in accurately forecasting future events. It is widely acknowledged that there exist significant pressures to formulate a novel trajectory. The issue of energy security has prompted the emergence of two contrasting approaches: the expansion of nuclear power generation and the promotion of renewable energy sources. The locus of the debate is primarily centered on Europe. As an illustration, the Swedish government has recently declared its intention to rescind a three-decade prohibition on advancing additional nuclear power capabilities. The European Union has declared the 20/20/20 objectives for the year 2020. The demonstration of clean coal technology, incorporating carbon capture technology, is being conducted in conjunction with exploring geological storage alternatives. The European Union (EU) continues endorsing carbon trading as a viable strategy for mitigating emissions. The election of a new President in the United States signifies the commencement of a fresh phase of active participation in the global discourse surrounding climate change. Moreover, it appears probable that the United States will implement a Cap-and-Trade system resembling the European Union Emissions Trading Scheme (EU ETS) (Bolla et al., 2010).

In conclusion, policymakers face significant pressures to redirect the course of energy system development toward a low-carbon pathway. However, there is currently limited evidence or indication of how this objective will be achieved. It is understood that the decisions made will significantly impact the long-term trajectory of energy development. As an illustration, the proposition to establish a coal-fired power station at Kingsnorth in the United Kingdom entails a minimum operational lifespan of 30 years. Without the development of efficient carbon capture and storage technology, this power station will continuously release carbon emissions into the atmosphere over its duration. Some proponents assert more extreme perspectives, contending that a comprehensive transformation of the energy economy is imperative, necessitating the establishment of a hydrogen-based energy economy. There exist compelling justifications to endorse the action above, as hydrogen combustion produces water, a benign compound. This chapter will examine the prospects of establishing a hydrogen-based energy economy. Additionally, this study will examine the potential effectiveness of carbon trading to reduce carbon emissions and promote technological advancements (Quader et al., 2016).

A relatively overlooked facet of the discourse surrounding energy pertains to human behavior. It is contended that the fundamental basis for any strategy for advancing energy systems should prioritize optimizing efficiency in both demand and supply aspects. The efficiency of the supply side is typically influenced by technological advancements that arise as a result of policy modifications or emerging factors, such as the introduction of biofuels. The demand side is also influenced by technology and new developments. The concept of lifestyles, encompassing our perspectives and utilization of energy services, has emerged as a crucial determinant, highlighting the substantial influence of social learning (Blaabjerg & Ionel, 2015).

## 8.1 HYDROGEN AND THE ENERGY SYSTEM

Hydrogen is a viable energy resource when it exists in its unbound form. When oxygen is combined with it, energy is released in the form of heat. Water is the resultant byproduct of the chemical above reaction. Hydrogen possesses the capacity to generate energy devoid of carbon emissions. There is a prevailing belief among certain individuals that hydrogen has the potential to serve as the primary fuel source in the future. Rifkin presents a vision of a future society where hydrogen emerges as the prevailing energy source (Rosen & Koohi-Fayegh, 2016).

Nevertheless, a significant drawback associated with a hydrogen energy economy is that, despite its evident environmental advantages, hydrogen is not naturally abundant in its elemental state and necessitates production processes. This necessitates the expenditure of energy. The process of electrolysis can be utilized to generate hydrogen from water. During this procedure, electrical energy is applied to water, separating its constituent elements, namely hydrogen and oxygen. Hydrogen serves as an efficient carrier of energy within the process. To clarify, hydrogen serves as a medium for energy transfer. The process of combining hydrogen with oxygen results in the release of energy and the formation of water. This phenomenon can be described as a self-sustaining cycle in which hydrogen production occurs, and its subsequent recombination yields beneficial energy without generating any detrimental byproducts. Nevertheless, the energy required for the process must be derived from a designated energy source. If fossil fuels were employed in the power generation process for electrolysis, it would result in the emission of harmful substances, thereby diminishing the prospective advantages associated with this particular procedure (Midilli et al., 2005).

**Did You Know?**

Hydrogen, being the least dense and most prevalent chemical element in the cosmos, constitutes approximately 75% of the total elemental mass in the universe. The significant quantity of this resource enhances its capacity to serve as a valuable medium for energy storage and transportation.

An alternative approach to hydrogen production entails the process of natural gas reforming, which entails the removal of the carbon atom. The resulting compound is carbon dioxide. Electrolysis and reformation represent the prevailing techniques employed for hydrogen generation; however, both methods exhibit certain limitations (Eriksson & Gray, 2017). The reformation process necessitates using carbon dioxide storage techniques, such as Carbon Capture and Storage (CCS). In order to implement electrolysis effectively, the utilization of carbon capture and storage (CCS) technology or the adoption of renewable or nuclear energy sources for electricity generation would be necessary. As elucidated subsequently within this section, employing electricity produced from renewable sources directly, rather than utilizing it for hydrogen production, can yield greater efficiency. Additional techniques for hydrogen production encompass coal gasification, necessitating carbon capture, and storage (CCS), biological production via a bioreactor, as well as thermal decomposition or thermolysis (Sørensen et al., 2004).

Hydrogen exhibits versatility as a fuel source. As an illustration, it can be employed within a combustion process to generate thermal energy and mechanical output or in conjunction with a fuel cell to generate electrical energy. The diagram presented in Figure 8.1 illustrates the diverse functions of hydrogen within global energy frameworks (Petkov & Gabrielli, 2020).

The utilization of hydrogen and fuel cells in transportation systems has garnered significant attention due to their potential to provide environmentally friendly modes of transportation, yielding substantial advantages for urban regions. While there are notable environmental advantages associated with decreased greenhouse gas emissions and urban pollution, concerns arise regarding the effectiveness of the process due to the energy requirements for storing hydrogen in a manner suitable for mobile applications. In Figure 8.2, the depicted data represents the number of steps necessary, each necessitating an energy input, in order to achieve the adequate compression or liquefaction of hydrogen to store it within a vehicle's fuel tank. The left-hand side of Figure 8.2 illustrates this. Every individual stage involved in the process of preparing hydrogen for onboard storage necessitates an input of energy, thereby resulting in a reduction of the overall cycle's efficiency. The route depicted on the right side of Figure 8.2 illustrates the utilization of electricity as a direct power source for a vehicle. Once again, it should be noted that there are losses incurred during the process of battery charging. However, it is important to acknowledge that the overall loss in this context is comparatively lower than that observed in the hydrogen route (Dawood et al., 2020).

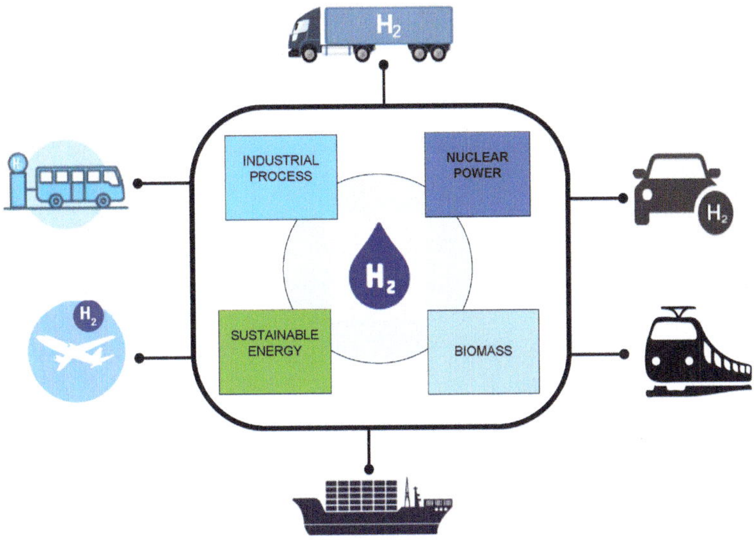


**Figure. 8.1.** Role of hydrogen in global energy scenarios (Source: Dash SK, Chakraborty S, Roccotelli M, Sahu UK. Hydrogen Fuel for Future Mobility: Challenges and Future Aspects. Sustainability. 2022; 14(14):8285. https://doi.org/10.3390/su14148285, Creative Commons License).

The analysis presented in Figure 8.2 solely focuses on a singular facet within the context of a prospective hydrogen economy while neglecting to address the crucial matter of vehicle range. If battery and hydrogen storage technologies achieve ranges comparable to or surpassing those of conventional vehicles in the future, the utilization of battery technology would be deemed the most efficient approach.

While there have been advancements in the range of electric and hydrogen-powered vehicles, further progress is necessary to achieve parity with hydrocarbon-based vehicles. Furthermore, there exist concerns about the storage of hydrogen on board. Currently, high-pressure or cryogenic techniques demonstrate the most favorable prospects. However, addressing the safety concerns associated with these methods is imperative. Nevertheless, as previously mentioned, substantial capital will be required to develop infrastructure (Tamalouzt et al., 2016).

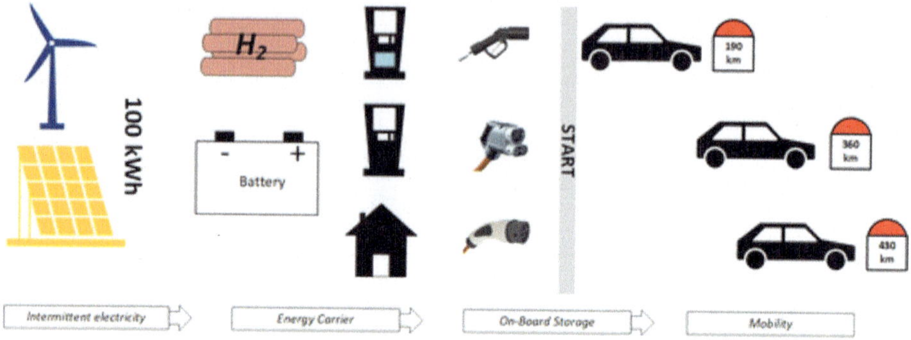


**Figure. 8.2.** Representation of hydrogen route versus electrical route for vehicles (Source: Ligen Y, Vrubel H, Girault HH. Mobility from Renewable Electricity: Infrastructure Comparison for Battery and Hydrogen Fuel Cell Vehicles. World Electric Vehicle Journal. 2018; 9(1):3. https://doi.org/10.3390/wevj9010003, Creative Commons License).

The illustration about the automotive industry prompts inquiries of a more comprehensive nature regarding the viability of an energy system reliant on hydrogen. It is evident that when there is abundant renewable energy capacity and a corresponding demand, it is generally more efficient to utilize the available power directly. The production and storage of hydrogen can be facilitated, allowing for its utilization later in time, given the existence of a mechanism for harnessing renewable resources. It is crucial to acknowledge and incorporate the supplementary energy expenditure associated with this process into the development of the overall system. The utilization of hydrogen derived from surplus renewable energy capacity has the potential to facilitate power generation. There exist instances wherein the electron economy may be deemed a more favorable alternative than the hydrogen economy and, conversely, situations where the hydrogen economy may be preferred over the electron economy. While concerns persist regarding fuel cells and hydrogen application in the automotive industry, these technologies are employed to generate electricity for residential and commercial use. In this particular case, the imperative to reduce the dimensions and mass of the equipment is not as urgent (Chapman et al., 2019).

As previously mentioned, fuel cells are modular, allowing customization to accommodate specific electrical demands. These structures are commonly referred to as fuel cell plants. Fuel cell plants possess several advantages when employed for on-site power generation, including reduced emissions of pollutants. These efficiencies are comparable to thermal generators and a near absence of noise. Fuel cell plants have been designed and implemented to supply heat and power to residential properties. Commercial systems utilizing natural gas as a fuel source have been designed and manufactured with power capacities spanning from 1 to 4.5 megawatts (MW). The electrical efficiency of these systems is approximately 40%, a level comparable to that of thermal plants. Utilizing the heat produced by the fuel cell enables the system to achieve an impressive overall efficiency of up to 80 percent. If this technology were integrated with hydrogen production methods that employed renewable, nuclear, or fossil fuel systems with carbon capture and storage (CCS), such an approach would offer a distinct environmentally friendly niche for fuel cell plants (Yue et al., 2021).

**Did You Know?**

Hydrogen fuel cells are extensively utilized in remote areas and off-grid scenarios to furnish portable energy for diverse objectives.

This particular arrangement provides insight into the potential efficacy of hydrogen in the context of the renewable energy sector. The optimal implementation of renewable energy sources is most effectively achieved locally, necessitating a development strategy that adheres to the principles of "capture/harvest-when-available" and "store until required." While it is possible to make general assumptions about the availability of resources, it is not feasible to accurately match the resource to real-time demand in an energy system that relies on intermittent supplies. This suggests that the configuration of a renewable energy economy is geographically limited and dependent on the available local resources. Figure 8.3 presents a comprehensive depiction of a renewable energy economy, highlighting various facets of the hydrogen economy. It should be noted that Figure 8.3 illustrates the case of the city. Nevertheless, it is conceivable to devise a system that incorporates thermal energy (Afgan & Carvalho, 2004).

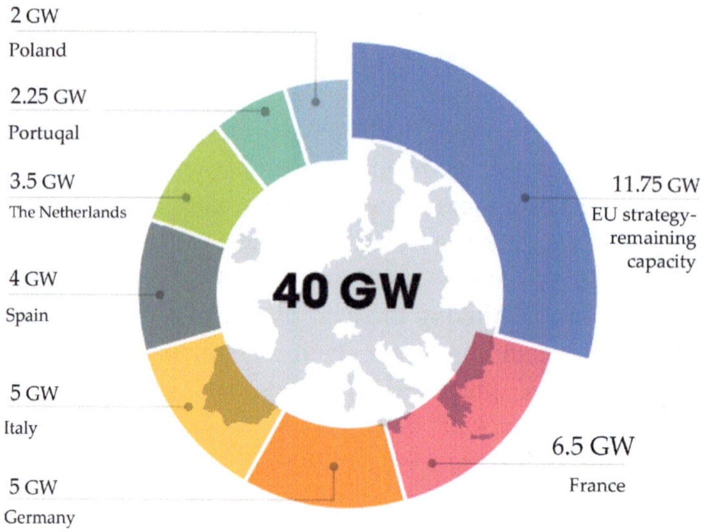


**Figure. 8.3.** Illustration of the hydrogen economy (Source: Sadik-Zada ER. Political Economy of Green Hydrogen Rollout: A Global Perspective. Sustainability. 2021; 13(23):13464. https://doi.org/10.3390/su132313464, Creative Commons License).

The utilization of various renewable technologies, such as wind, solar, photovoltaic, wave, tidal, and biomass, for harvesting is promising. If renewable capacity surpasses demand, the surplus energy, encompassing both heat and power, can be utilized to produce hydrogen as a type of energy storage. In instances where demand surpasses supply, hydrogen can be utilized to produce both electrical power and heat. Additionally, it can be utilized for practical applications such as transportation. The system's administration would be conducted by an intelligent grid system overseeing the allocating and utilization of various resources. The scenario above highlights the potential contribution of hydrogen in a sustainable energy sector. Efforts have been made to integrate various renewable electricity generation technologies and effectively manage them using an intelligent electrical grid system. Currently, these technologies are being implemented on a limited scale, although the possibility of expanding their application exists. This intelligent management approach can also be extended to heat networks (Parra et al., 2019).

Nevertheless, the development of such systems in a finalized state is improbable. The anticipated developmental trajectory is expected to commence with smaller systems with the capacity for evolution. The level of familiarity and proficiency in utilizing renewable technologies varies among individuals. The underlying premise of the current system is that the user assumes a primarily passive role, relying on readily available energy services as needed. Based on the available evidence, it is evident that there is a need for increased participation in the process of system development. The analysis of small-scale renewable projects reveals several key factors contributing to their successful implementation and advancement. These findings can be briefly summarized as follows:

- Conducting a needs assessment to ensure a comprehensive understanding of energy requirements is established.
- Energy mapping for the identification and assessment of local energy resources.
- The requisite support systems encompass technical, human, and financial aspects.
- The appropriate level, including the establishment of the entry-level.

Learning by builders and users is a fundamental aspect of achieving success in system development. In the context of technology transfer, avoiding making assumptions regarding a singular direction is imperative. System developers must ensure the transferred technologies are suitable for the user's capacity and the available resources. Research on localized solutions demonstrates the feasibility of smaller intelligent systems and micro-grid technological approaches. These approaches enable a system to adapt and improve as users become more comfortable with the technology and their capacity to manage the system effectively. Hydrogen possesses the potential to generate environmentally friendly energy. However, there are associated expenses with this. When considering a comprehensive strategy for developing future energy systems, it becomes evident that hydrogen has a significant role to fulfill. Nevertheless, due to certain inefficiencies associated with current hydrogen production technologies, the initial deployment of hydrogen will probably be most effectively utilized in specific and specialized applications. If a viable and secure approach to onboard storage can be discovered, hydrogen could potentially play a substantial role within the transportation industry. In long-term system development, hydrogen has the potential to assume a significant role. However, it should be noted that hydrogen would not be the system's sole focus but rather one of several strategies aimed at guaranteeing sustained access to a diverse array of energy services (Uyar & Beşikci, 2017).

## 8.2 MARKET MECHANISMS AND DE-CARBONIZING THE ENERGY SYSTEM

The Kyoto Protocol implemented a market-oriented strategy for mitigating greenhouse gas emissions. The rationale underlying implementing a Cap-and-Trade system is predicated on the notion that it will catalyze fostering innovation. This system posits that companies capable of devising novel low-carbon technologies will assume positions of market leadership, thereby reaping the rewards of not only generating revenue through the sale of their products but also through the sale of surplus carbon credits (Vourdoubas, 2021). At first glance, this proposition appears straightforward, resulting in benefits for all parties involved, including the environment. The contention posits that trade catalyzes innovation, ultimately resulting in substantial reductions in carbon emissions and the gradual establishment of low-carbon energy systems. This section aims to assess the efficacy of carbon trading, given its apparent status as the sole available option. Currently, the EU Emissions Trading Scheme (EU ETS) is the most extensive carbon trading system in operation, primarily confined to the European region. It has become evident that the United States administration also demonstrates a preference for implementing a Cap-and-Trade system. This implies that forthcoming energy strategies will incorporate a substantial element of Cap and Trade within their framework (Varanasi & Tripathi, 2020).

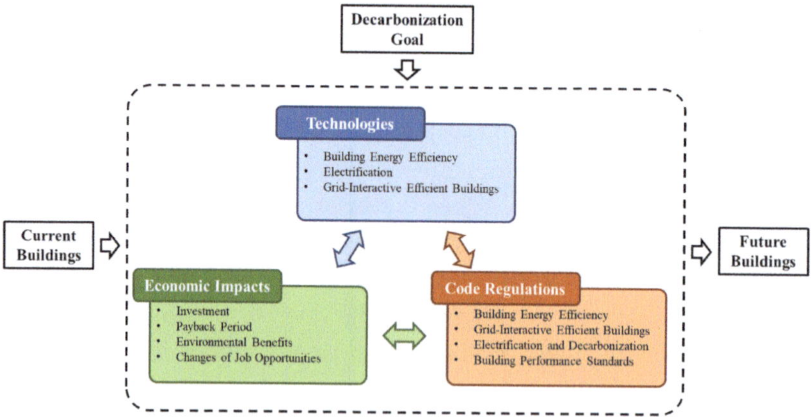


**Figure. 8.4.** Illustration of decarbonizing the energy system (Source: Ye Y, Dehwah AHA, Faulkner CA, Sathyanarayanan H, Lei X. A Perspective of Decarbonization Pathways in Future Buildings in the United States. Buildings. 2023; 13(4):1003. https://doi.org/10.3390/buildings13041003, Creative Commons License).

## 8.3 EMISSION TRADING

Emission trading is a regulatory strategy aimed at mitigating pollution by offering economic incentives for the reduction of emissions. The US Environmental Protection Agency implemented the inaugural emissions trading program in 1976. The policy enabled the construction of additional environmentally harmful facilities in return for implementing 'offsets' that effectively mitigated air pollution to a greater extent compared to other sources within the corresponding area. The perspective that markets have the potential to offer more economically efficient approaches to addressing pollution issues compared to regulatory measures reached its pinnacle with the enactment of the Clean Air Act Amendments of 1990. This legislation established a nationwide program for trading sulfur dioxide emissions, enabling power plants to save costs while effectively managing acid rain. Additionally, it incentivized states to adopt emissions trading to mitigate urban smog. Two market mechanisms that have emerged from this experience are commonly referred to as Cap and Trade and Project Based or Offset mechanisms (Zhang et al., 2020).

**Cap and Trade** is an effective policy strategy for managing and reducing significant emissions from a collective set of sources, offering cost advantages compared to individual source regulation. The proposed methodology initially establishes a comprehensive limit, referred to as an overall cap, which represents the uppermost threshold of emissions permissible within each compliance period in order to attain the intended environmental outcomes. The allocation of emission allowances to affected sources is a means of granting permissions to emit, with the condition that the cumulative number of allowances does not surpass the predetermined limit. Over the course of time, there is a gradual reduction in the level of the cap (Stuhlmacher et al., 2019).

**The Project-Based or Offsets** approach entails the acquisition of emission credits through the implementation of projects that generate reductions in emissions that would not have materialized in the absence of said projects. The mechanisms above have emerged as the focal point of the market-oriented strategy for mitigating greenhouse gas (GHG) emissions within the framework of the Kyoto Protocol. The establishment and operation of carbon markets and trading mechanisms have emerged as a progressively significant component in addressing the challenge of mitigating carbon emissions (Ohndorf, 2010).

**Did You Know?**

The inaugural carbon capture and storage facility of significant magnitude, known as "Sleipner," has been in operation within Norway since 1996. The process of capturing and storing carbon dioxide from natural gas production has resulted in the accumulation of more than 23 million metric tons of this greenhouse gas.

## 8.4 CARBON TRADING

The concept of carbon trading entails a financial exchange in which one party compensates another party in exchange for greenhouse gas emission assets, which the purchaser can utilize to fulfill their objectives. There exist two primary classifications of transactions (Pan & Dong, 2023).

The concept of allowance-based transactions involves purchasing emission allowances issued by regulatory bodies within cap-and-trade systems. Examples of such allowances include Assigned Amount Units (AAUs) under the Kyoto Protocol and European Union Allowances (EUAs) under the European Union Emissions Trading Scheme (EU ETS). These schemes integrate environmental performance, which is determined by the regulator's issuance of allowances to limit global emissions from mandated entities, with flexibility through trading. This allows participants to fulfill compliance requirements in the most cost-effective manner (Song et al., 2022).

Project-based transactions involve buyers acquiring emission reductions (ERs) from projects that can provide verifiable evidence of their ability to reduce greenhouse gas (GHG) emissions compared to alternative scenarios. Examples of such projects include investing in wind power or other renewable energy sources rather than coal-fired power generation or implementing energy efficiency measures at large industrial facilities to decrease energy demand and consequently reduce GHG emissions from power generation. The CDM (Clean Development Mechanism) and JI (Joint Implementation) mechanisms of the Kyoto Protocol are notable instances of such activities, which result in the production of Certified Emission Reductions (CERs) and Emission Reduction Units (ERUs), respectively. These mechanisms are commonly referred to as the Flexibility Mechanisms (Fang et al., 2020).

**Did You Know?**

The introduction of carbon trading emerged as a pivotal mechanism within the framework of the Kyoto Protocol in 1997.

**Table 8.1.** Carbon markets: Volumes and values 2006–2007 (Source: Phil O' Keefe, Creative Commons License).

| | | 2006 | | | 2007 | |
|---|---|---|---|---|---|---|
| | **Volume (MtC O$_2$ O$_2$)e** | | **Value (MUS$)** | **Volume (MtC O$_2$ O$_2$e)** | | **Value (MUS$)** |
| Allowances | | | | | | |
| EU ETS | 1,104 | | 24,436 | 2,061 | | 50,097 |
| New South Wales | 20 | | 225 | 25 | | 224 |
| Chicago Climate | 10 | | 38 | 23 | | 72 |
| Subtotal | 1,134 | | 24,699 | 2,109 | | 50,394 |
| Project-based | | | | | | |
| Primary CDMa | 537 | | 5,804 | 551 | | 7,426 |
| Secondary CDM | 25 | | 445 | 240 | | 5,451 |
| JI | 16 | | 141 | 41 | | 499 |
| Other compliance | 33 | | 146 | 42 | | 265 |
| Subtotal | 611 | | 6,536 | 874 | | 13,641 |
| Total | 1,745 | | 31,235 | 2,983 | | 64,035 |

Carbon trading should be regarded as one among several strategies for mitigating greenhouse gas (GHG) emissions, including enhancing efficiency and fostering the growth of renewable energy sources. Implementing flexibility mechanisms has resulted in a significant expansion of carbon trading, leading to the establishment of multiple markets. Based on data provided by the World Bank, the estimated worth of carbon trading in 2007 amounted to approximately US$64 billion, representing a growth rate of slightly more than twofold compared to the preceding year, 2006. The European Union Emissions Trading System (EU ETS) emerged as the most substantial individual market, boasting a trade volume of approximately US$50 billion, which accounted for nearly 80 percent of the worldwide market (Hua et al., 2020).

According to estimates, approximately 4.1 billion metric tonnes of carbon dioxide equivalents are annually introduced into the Earth's atmosphere. According to the data presented in Table 8.3, the trading of carbon credits in 2007 reached a significant volume of nearly 3 billion tonnes of carbon dioxide equivalents. This observation implies that adopting this approach holds significant promise for mitigating greenhouse gas emissions. Nevertheless, the actuality diverges significantly (Long et al., 2022).

## 8.5 EFFECTIVENESS OF CARBON TRADING IN MITIGATION

One of the fundamental principles underlying trading is the belief that market mechanisms can identify and implement the most efficient and economically viable solution to a given problem. Successfully mitigating greenhouse gases necessitates a decrease in the activities that generate said gases. Significant reductions on a large scale necessitate the implementation of substantial changes. To mitigate greenhouse gas emissions, it is imperative to transition towards energy systems that prioritize the utilization of efficient and low-carbon technologies. The climate issue encompasses a significant social aspect alongside notable technological dimensions that contribute to potential solutions. Nevertheless, empirical evidence derived from the US Sulphur Dioxide program's implementation of Cap and Trade indicates that the market-based approach does not effectively stimulate innovation. This implies that regulatory measures might serve as a more efficacious catalyst for technological advancements (Gao et al., 2020).

This matter holds significant importance, especially when contemplating the largest contributors to carbon emissions, namely the energy corporations. Cap and Trade fails to consider the specific characteristics and nuances of different industries. Numerous prominent electricity producers have made substantial investments in fossil fuel infrastructure, resulting in their entrenchment within a fossil fuel-based economic framework. Instead of transitioning to renewable energy sources, it is probable that there will be a proclivity towards pursuing enhancements in efficiency (considering the advanced stage of the technology, these improvements will only be incremental) and procuring emission credits, with the resulting costs being transferred to consumers. Cap and Trade offer incentives for industries that produce greenhouse gases but are not inherently dependent on fossil fuels. Nevertheless, in the short term, it is more probable for these industries to prioritize immediate gains by trading their emission rights rather than investing in low-carbon solutions that would effectively reduce their emissions and yield marketable technologies. In summary, the implementation of Cap and Trade is more inclined to incentivize the prioritization of cost-effective approaches for reducing emissions rather than promoting the transformative changes that have been advocated as essential for mitigating the risks associated with climate change (Li & Jia, 2017).

The construction of the carbon market has engendered additional challenges. This phenomenon is particularly conspicuous within the EU ETS, which stands as the largest carbon market. The initiative commenced in the year 2005 and encompassed approximately 10,000 energy-intensive facilities within the European Union, which collectively accounted for approximately 40% of carbon emissions. The allocation of allowances, amounting to a value of €120 billion, was distributed free of charge in most member states, except four states where a portion of allowances were auctioned. However, it is worth noting that only one member state fully auctioned 5 percent of allowances, as permitted during the initial phase. Each member state determined these allocations through National Allocations. Many of these targets were established at excessively high levels, resulting in a decline in carbon prices upon the commencement of market operations in 2005. The industries encompassed within the European Union Emissions Trading System (EU ETS) emitted 66 million tonnes below the predetermined cap level (Cheng et al., 2016).

The existence of generous allowance levels implemented by certain member states, undermines the principle of scarcity, which is crucial for maintaining a strong market, and raises concerns about the commitment of EU members to fulfill their international climate obligations. According to Grubb (year), it is observed that:

Suppose the existing national allocation plans are permitted to remain in place. In that case, there is a significant risk of undermining the trustworthiness of the EU ETS and the efficacy of carbon trading as a means to address carbon emissions (Tang et al., 2019).

The initiation of the second phase of the Emissions Trading Scheme (ETS) occurred in January 2008. During this phase, the European Union expanded the range of activities eligible for auctions. However, it is worth noting that out of all the EU members, only ten chose to participate in this initiative. Furthermore, it is important to highlight that four of these ten members allocated less than 1 percent of their total allocations through the auction process. According to Neuhoff's argument in 2006, the proposed levels of subsidies in the second phase could potentially result in distorted allocation decisions. This could potentially create a scenario where constructing a coal-fired power station would be more financially advantageous than when the Emissions Trading Scheme (ETS) was not in place. Despite the implementation of stricter caps in this particular phase, detractors of the system contend that the Linking Directive, which was introduced in 2004 and put into effect in 2005 and serves as a connection between the EU ETS and the flexibility mechanisms outlined in the Kyoto Protocol, undermines the cap. This is due to the fact that it permits the utilization of credits obtained from Clean Development Mechanism (CDM) and Joint Implementation (JI) projects to offset domestic targets (Wang et al., 2015).

Furthermore, critics argue that this phenomenon weakens the effectiveness of domestic climate policies implemented by member states and has a detrimental impact on the European Union's credibility during climate negotiations. Nevertheless, the expeditious and effortless implementation of this measure exemplifies the considerable sway wielded by member states during EU negotiation procedures, as well as their inclination towards establishing a framework that alleviates strain on domestic endeavors. The initial perspective on the European Union (EU) and its institutions emphasized the avoidance of blending market mechanisms, but this stance has evolved to the extent that market mechanisms are now increasingly perceived as a more efficient approach for achieving environmental objectives (Anke et al., 2020).

This observation signifies the progressive transition from a command-and-control approach to the adoption of market-based alternatives in the formulation of environmental policies within the European Union. Research conducted by Point Carbon Advisory Services, which was commissioned by the World-Wide Fund for Nature (WWF), investigated the potential windfall profits that could be obtained during Phase 2 of the European Union Emissions Trading Scheme (EU ETS). The study focused on five EU member states, namely Germany, the United Kingdom, Italy, Spain, and Poland. The findings of the study indicate that the total windfall profits could range from 23 to 71 billion, depending on the price of European Union Allowances (EUAs). The potential for obtaining profits without any effort may also contribute to understanding the relative ease with which the second phase of the scheme was approved (Paul et al., 2023). In 2008, the European Commission put forth several proposed modifications to the scheme. These changes encompassed the elimination of national allocation plans in favor of centralized allocation by a European Union (EU) authority. Additionally, there was a shift towards auctioning a larger proportion (60% or more) of permits instead of distributing them freely. Furthermore, the proposed amendments sought to incorporate nitrous oxide and perfluorocarbons, which are greenhouse gases (European Commission, 2008).

The European Union (EU) has set a target to limit industrial emissions in the year 2020 to a maximum of 21 percent below the levels recorded in 2005. This equates to a total of 1720 million allowances for emissions. In order to ascertain the aggregate emissions, there will be an annual reduction of 1.74 percent in the number of allowances in circulation by the end of 2012.

- The scheme is proposed to undergo expansion to encompass additional sectors such as aviation, petrochemicals, ammonia, and the aluminum sector. Furthermore, the inclusion of two new gases, namely nitrous oxide, and perfluorocarbons, would result in the coverage of approximately 50% of all emissions within the European Union. Road transport and shipping are currently not included, although shipping may be considered for inclusion in the future. The exclusion of agriculture and forestry from the analysis is attributed to the challenges associated with accurately quantifying emissions originating from these sectors (Golden et al., 2011).
- In order to attain a mean reduction of 10 percent in greenhouse gas emissions from sectors not encompassed by the Emissions Trading System (ETS), including transportation, buildings, agriculture, and waste, by the year 2020, the Commission has established individualized national targets based on countries' gross domestic product (GDP). We observe that wealthier nations are being urged to implement more substantial reductions in greenhouse gas emissions, with Denmark, Ireland, and Luxembourg facing reductions of up to 20 percent. Conversely, less affluent countries, particularly Portugal and those that joined the European Union after 2004, except Cyprus, will be permitted to increase their emissions in these sectors. Romania and Bulgaria, for instance, may increase their emissions by up to 19 and 20 percent, respectively, aligning with their ambitious GDP growth targets (Ke et al., 2023).
- Smaller installations with an annual $CO_2$ emission of less than

10,000 tonnes will have the option to exempt themselves from the Emissions Trading Scheme (ETS) on the condition that they implement alternative measures to reduce their emissions. Using carbon capture and storage (CCS) technology effectively prevents the release of industrial greenhouse gases into the atmosphere. Consequently, these gases can be acknowledged as having not been emitted within the Emissions Trading System (ETS) framework (Shojaei & Mokhtar, 2022).

- Currently, pollution allowances predominantly involve the gratuitous distribution of 90 percent to industrial installations. However, there are imminent plans to significantly augment the practice of auctioning, with implementation expected to commence as early as 2013. The projected figure suggests that approximately 60 percent of the overall quantity of allowances will be subjected to auction in the year 2013. It is anticipated that complete auctioning for the power sector will become the prevailing practice starting in 2013. This shift is projected to result in a 10-15 percent increase in electricity prices. In various industries, the process of gradually eliminating free allocations will occur annually from 2013 to 2020. However, specific sectors that are highly energy intensive may continue to receive their allowances free of charge in the long run, provided that the Commission identifies them as being at substantial risk of carbon leakage, which refers to the relocation of these sectors to third countries with less stringent climate protection regulations. The determination of the sectors impacted by this measure remains pending (Dhanda et al., 2022).
- The development of the distribution method for free allowances will be undertaken by expert panels within the Commission at a subsequent stage. The basis for these criteria is likely to be derived from grandfathering or historical emissions and performance considerations.
- In order to mitigate the potential negative impact on European competitiveness resulting from carbon leakage, which refers to external industries exporting to the EU and gaining a competitive advantage due to less stringent climate regulations, the European Union (EU) will contemplate implementing certain compensatory measures to safeguard its domestic industries, should a global agreement fail to materialize.

- The Role of Flexibility and Third Countries in Achieving Climate Change Goals: If a comprehensive international agreement on climate change is achieved, member states will retain the option to fulfill a portion of their emission reduction targets by providing financial support for projects aimed at reducing emissions in countries outside the European Union. However, the utilization of such credits will be restricted to a maximum of 3 percent of member states' total emissions in the year 2005, which corresponds to approximately one-quarter of the overall reduction endeavor.

Currently, the proposals above are subject to negotiation within the European Union, and they will probably undergo significant modifications compared to the propositions above. Assessing the future effectiveness of Emissions Trading Schemes (ETS) in mitigating greenhouse gas (GHG) emissions is currently challenging (Cadez & Czerny, 2016).

## 8.6 CDM AND JI

There is a prevailing concern regarding the potential utilization of credits obtained through these mechanisms to counterbalance domestic targets, thereby compromising the ability to instigate substantial reform within sponsoring entities. In order for a JI project to be deemed acceptable, it is imperative that it obtains the consent of all parties involved and demonstrates a quantifiable decrease in emissions from sources or an increase in removals by sinks that is supplementary to any reductions or enhancements that would naturally transpire. The Joint Implementation (JI) mechanism is subject to the oversight of a governing body known as the "Supervisory Committee." Similar to the Clean Development Mechanism (CDM), the certification of Emission Reduction Units (ERUs) within JI requires the involvement of "Independent Entities." The accrual of JI Emission Reduction Units is limited to the period after 2008. Nevertheless, certain countries have encouraged investments in early JI emission reduction projects and have allocated post-2008 emission rights from their budgets to account for the reductions made before 2008 (Patlitzianas et al., 2007).

Both the Joint Implementation (JI) and Clean Development Mechanism (CDM) have been subject to criticism. The mechanism of Joint Implementation (JI) facilitates the buying and selling of emissions credits among countries, potentially providing an avenue for countries to fulfill their Kyoto Protocol obligations through financial means. When prioritizing cost-effective projects, it is advisable to assign a lower priority to more expensive renewable energy and energy efficiency initiatives, as they tend to generate carbon credits at a higher cost. Leakage represents an additional area of concern. Leakage pertains to external ramifications, such as the scenario where a project aimed at enhancing the efficiency of automobiles results in reduced emissions. However, this reduction is achieved indirectly, as the project's cost reduction in driving may potentially incentivize car owners to amplify their vehicular usage and potentially acquire additional cars (Woerdman, 2000).

## 8.7 THE CLEAN DEVELOPMENT MECHANISM (CDM)

The Clean Development Mechanism (CDM) enables Annex I Parties, which consist of developed industrialized countries that are members of the OECD and EIT (economies in transition), to undertake project activities aimed at mitigating emissions in non-Annex I countries, which primarily comprise the developing world. The approval of all Parties involved is a prerequisite for CDM project activities, which are required to achieve emission reductions below the baseline emissions that would have taken place in the absence of the CDM project activity. The Marrakech Accords established a comprehensive project cycle for the Clean Development Mechanism (CDM), which is supervised by the CDM executive board (EB). The board consists of ten members who are elected by the Conference of the Parties (COP) of the United Nations Framework Convention on Climate Change (UNFCCC). The Board is responsible for registering projects and verifying their compliance with the established regulations. The simplified regulations for small-scale Clean Development Mechanism (CDM) projects were established during the eighth Conference of the Parties (COP 8). In contrast, the regulations for sequestration CDM projects were agreed upon during the ninth Conference of the Parties (COP 9) (Subbarao & Lloyd, 2011).

The inclusion of CDM emission credits in the aggregate emissions budget of certain nations necessitates the assurance of their quality. Therefore, the accumulation of emission credits occurs exclusively following independent verification conducted by "Operational Entities" (OEs), primarily consisting of commercial certification firms. The emission credits are commonly known as Certified Emission Reductions (CERs) (Sutter & Parreño, 2007). Within the context of Clean Development Mechanism (CDM) projects, it is a requirement that a portion of the generated proceeds be allocated toward the advancement of sustainable development initiatives within the host country. Critics contend that many projects fail to effectively tackle the underlying issue of greenhouse gas (GHG) reduction in industrialized nations while also neglecting to foster sustainable development within the host country. CDM encompasses a diverse array of projects, varying in terms of type and scale. Several studies have primarily concentrated on the development of large-scale hydropower projects and the implementation of clean coal initiatives. Projects of such magnitude have the potential to yield substantial credits and are likely to garner interest from investors. However, it is worth noting that these projects may not necessarily foster sustainable development. The 2000 report published by the World Commission on Dams presented evidence indicating that extensive hydropower initiatives have had significant adverse effects on social and environmental aspects and have consistently failed to meet expected performance standards (Lim & Lam, 2014).

Furthermore, it can be argued that the implementation of CDM clean coal projects fails to address future resource concerns effectively and poses obstacles to the advancement of renewable energy initiatives, ultimately impeding the promotion of sustainable development. There is a prevailing belief among scholars and experts that the Clean Development Mechanism (CDM) may not be a productive instrument for facilitating the advancement of renewable energy initiatives in developing nations. This skepticism stems from the inherent structure of the CDM, which prioritizes the pursuit of carbon credits with the lowest possible cost. Expensive projects such as renewable energy initiatives are often marginalized due to the lack of recognition and reward for their numerous benefits (Pearson, 2007).

The CDM has demonstrated limited effectiveness in facilitating initiatives that target energy efficiency and transportation, both pivotal in advancing sustainable development in the global South and mitigating climate change on a global scale. According to a report by the World Bank, there is a substantial potential for efficiency projects. However, the small number of projects implemented thus far indicates that there are barriers that have not been fully accounted for in analyses of the potential for these projects. Additionally, research conducted by the Organisation for Economic Co-operation and Development (OECD) revealed that a significant and swiftly expanding segment of the Clean Development Mechanism (CDM) project portfolio exhibits limited direct impacts on the environment, economy, or society, apart from greenhouse gas (GHG) mitigation. Moreover, these projects generate minimal outcomes other than emissions credits (Dechezleprêtre et al., 2008). The World Bank asserts that CDM projects have the potential to facilitate sustainable development, albeit acknowledging the limited empirical substantiation available to substantiate this perspective. According to the Bank, there is evidence of a developing market niche that specifically recognizes and rewards CDM projects that effectively provide significant sustainability benefits. In summary, the available evidence regarding the efficacy of carbon markets in mitigating greenhouse gas (GHG) emissions is limited. However, a preliminary examination conducted by Ellerman and Buchner on the European Union Emissions Trading Scheme (EU ETS) indicates that the market has reduced emissions by approximately 2.5-5% compared to the projected levels in the absence of such a mechanism (De Coninck et al., 2007).

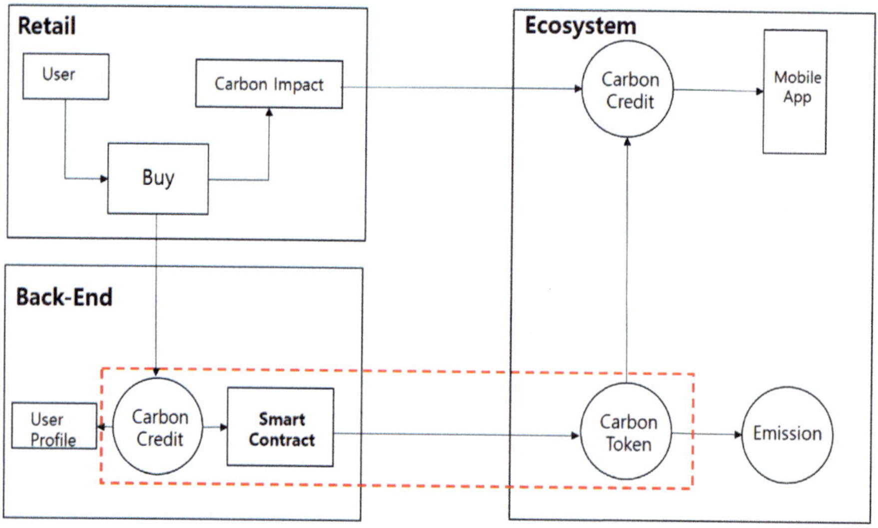


**Figure. 8.5.** Illustration of the carbon market (Source: Kim S-K, Huh J-H. Blockchain of Carbon Trading for UN Sustainable Development Goals. Sustainability. 2020; 12(10):4021. https://doi.org/10.3390/su12104021, Creative Commons License).

The potential of carbon trading to make substantial contributions towards achieving the necessary deep cuts in the long run, is yet to be determined. The World Bank report, State and Trends of the Carbon Market, 2008, highlights a concerning trend: the market is progressing towards becoming a comprehensive system, as depicted in Figure 8.5. This evolution is characterized by the adoption of various features commonly observed in financial markets, including the monetization of carbon as a precursor to securitization, the utilization of derivatives, and the establishment of bond transactions tied to the projected price of carbon, among others. The occurrence of the credit crunch and subsequent bank failures in 2008 can be attributed, to a significant degree, to the improper utilization of financial instruments of this nature. The extent to which adequate regulation will be implemented to guarantee transparency in the carbon market and its continued emphasis on achieving tangible reductions is yet to be determined (Zainuddin et al., 2017).

One additional aspect to consider is that in 2009, the market value of carbon experienced a significant decline, reaching historically low levels. The aforementioned situation has prompted inquiries regarding the necessity of implementing a minimum price for carbon. This would entail strengthening a system that has failed to achieve carbon reduction targets. There is currently a growing discourse surrounding the necessity of implementing a carbon tax. Several prominent figures, including Yale economist William Nordhaus, renowned climate scientist Jim Hansen, and the chief executive officer of Exxon, have advocated for implementing a carbon tax. There is an ongoing debate regarding the efficacy of resource-based taxes, such as a carbon tax, compared to market mechanisms, with proponents contending that the former is more effective in stimulating innovation. Nevertheless, for resource-based taxes like a carbon tax to be effective, they must be implemented alongside other instruments such as policy, legislation, regulation, and efficiency standards (Cames et al., 2016).

## 8.8 OTHER CARBON REDUCTION APPROACHES

As elucidated in the preceding section, there remains uncertainty regarding the feasibility of developing a carbon market that can effectively yield tangible reductions in carbon emissions. Additional options to the United Nations Framework Convention on Climate Change (UNFCCC) and its Kyoto Protocol include the Contraction and Convergence proposal and the Greenhouse Development Rights Framework (Nealson & Myers, 1992).

The 'Contraction and Convergence' framework entails the implementation of a system whereby emission allowances are allocated on a per capita basis, with the ultimate goal of achieving equitable distribution of carbon allowances at comparable levels. The current framework is primarily grounded in national allowances, thus neglecting broader global concerns such as development. However, it aspires to advocate for equitable carbon allowances across all nations. On the contrary, the Greenhouse

Development Rights Framework posits that it is inequitable for individuals of lower socioeconomic status to bear the financial burden of addressing the environmental consequences resulting from the extravagant lifestyles of the affluent (Huisingh et al., 2015).

In order to address climate change, it is imperative to achieve a stabilization of greenhouse gas (GHG) concentrations in the atmosphere at a level approximately 80 percent lower than the present levels. Simultaneously, it is crucial to ensure that individuals living in poverty are granted the opportunity to pursue development. The foundation of this approach is rooted in the polluter pays principle, as opposed to the 'common but differentiated responsibilities' framework of the United Nations Framework Convention on Climate Change (UNFCCC). The Greenhouse Development Rights Framework suggests a threshold of US$9000 as the median global income level, below which individuals would be exempted from the obligation to reduce carbon emissions. In other words, they are engaged in survival rather than contributing to environmental contamination, thereby rendering them exempt from any associated responsibilities. According to the Greenhouse Development Rights Framework, the estimated cost to mitigate catastrophic climate change equitably ranges from 1 to 3 percent of global GDP. This estimation aligns closely with the findings of the Stern Report. Preliminary estimations indicate that the United States would assume approximately 33 percent of the burden, while the European Union would account for approximately 25 percent. Conversely, China and India would each contribute less than 1 percent. The proposal put forth by the Greenhouse Development Rights Framework, which advocates for a global framework, is considered highly radical due to its emphasis on incorporating the development agenda within the global commons issue (Morris et al., 2009).

The issue of global climate change necessitates attention at the global and macroeconomic levels. This implies establishing a connection between the aforementioned initiative and the global poverty alleviation program outlined in the Millennium Development Goals. The potential expenditure associated with this worldwide endeavor may amount to approximately 5.5 percent of the global gross domestic product (GDP). However, it is crucial to weigh this against the potential economic losses, estimated to range between 5 and 20 percent of global GDP, as outlined in the Stern Report (Dong et al., 2013). In order to attain a worldwide reduction of 80 percent in emissions and effectively stabilize the climate at a level below a 2°C increase, it is imperative to employ the strategy of technological 'leapfrogging'. The occurrence of leapfrogging is contingent upon the accessibility of relevant technologies to the most economically disadvantaged individuals. This phenomenon can only transpire if wealthier industrialized nations collectively diminish their structural dependence on fossil fuels (Marsden et al., 2014).

## 8.9 THE CHANGING ENERGY LANDSCAPE

The task of creating an energy system that aligns with climate targets, sustainable development objectives, and energy security concerns poses a significant and formidable undertaking. Throughout this literary work, we have examined a diverse range of technological advancements that possess the potential to enhance the effectiveness of both the supply and demand aspects of energy systems. Furthermore, an examination has been conducted on technologies that possess the capability to harness renewable resources. The process of decarbonizing the energy system poses significant challenges. For certain individuals, the persistent utilization of large-scale technologies, such as nuclear power and carbon capture and storage (CCS), presents a viable approach to achieving decarbonization on the supply side (Pasqualetti, 2012).

Nevertheless, nuclear technology exhibits a lack of adaptability, necessitating the identification of a viable solution for the unresolved issue of nuclear waste management. Carbon capture and storage (CCS) is currently in the developmental phase, which gives rise to concerns regarding the enduring security of carbon stored within geological formations. Some proponents endorse the adoption of renewable technologies, and it has been proposed that Europe could fulfill its energy needs by harnessing solar technologies implemented in the Sahara desert. Implementing such a scheme would face significant political obstacles that must be addressed (Yildiz, 2014). There is a general consensus among many individuals that enhancing efficiency on the demand side is of utmost importance. In previous chapters, there has been a notable opportunity identified for enhancing the efficiencies of buildings and appliances. The previously mentioned aspect holds significant importance and serves as a crucial initial measure in the process of decarbonizing the energy system. The subsequent inquiry pertains to the requisite approach for the supply side. In the context of renewable technologies, acknowledging the distinction between concentrated energy supplies, such as fossil fuels, and the dispersed nature of renewable resources, such as solar and wind, is crucial (Sharma et al., 2019).

Consequently, adopting a renewable approach necessitates the implementation of a strategy that involves capturing and storing energy until it is required, taking into account the intermittent nature of these renewable sources. The issue of storage is of paramount concern. The effective storage of electricity poses a challenge, necessitating the simultaneous development of a hydrogen economy and a renewable energy economy to optimize the potential for capture and storage (Onuchic et al., 1997).

Numerous stakeholders are advocating for immediate measures to mitigate carbon emissions, despite the absence of a consensus regarding the specific level of stabilization and the associated timeline. It is evident that regardless of the agreed target and timetable, the process of transitioning the energy system will require a substantial amount of time. There is apprehension regarding the potential long-term commitment associated with adopting specific technologies, such as nuclear power, which may result in a prolonged development trajectory (Zerta et al., 2008). Additionally, there is a potential for financial resources to be diverted from the advancement and implementation of alternative solutions. Furthermore, the issue of energy security has become a prominent focus, leading to increased emphasis on the utilization of domestic resources and the establishment of a varied energy portfolio. It is also evident that a significant technological advancement will be necessary, such as in the case of motor vehicles, where it is probable that hybrid models will initially gain prominence, followed by a potential dominance of either electric or fuel cell vehicles, or possibly a combination of both, within the fleet (de Boer & Zuidema, 2015). The implementation of solar thermal water heaters and photovoltaic panels on residential rooftops has been observed. The implementation of new regulations pertaining to zero-emission buildings will sustain the ongoing progress in this area. The potential ramifications of these alterations on the energy sector are expected to be significant. The prevailing energy systems typically exhibit a pyramidal structure. This system can be classified as one characterized by concentrated ownership and a large number of passive consumers. The sole interface between the consumer and the energy system is the provision of energy services and the subsequent billing by the utility company (Bryngelson et al., 1995).

Bills commonly provide details regarding the quantity and cost of energy units consumed without offering insights into the corresponding energy services. This lack of information pertains to various aspects, such as the utilization of space and water heating, cooking activities, as well as the consumption of electricity for lighting and operating appliances (Roland et al., 1999).

The implementation of distributed technologies is anticipated to have an impact on the physical environment, albeit in a nuanced manner. For instance, the incorporation of solar thermal collectors and photovoltaic panels on rooftops, the adoption of LED street lighting, and the distinct acoustic emissions produced by electric or hydrogen vehicles are among the subtle alterations that can be expected. A number of modifications will be imperceptible, such as the enhanced thermal efficiency of structures. The contrasting feature is the decentralized configuration of the energy system, as depicted in the right portion of Figure 8.5 (Valentin et al., 2016).

Feedin-Tariffs (FiTs) have become increasingly prevalent in the renewable energy sector. Net metering refers to the practice of allowing energy consumers to generate their own electricity and feed any excess energy back into the grid. This allows consumers to offset their energy consumption and potentially receive credits or payments for the surplus energy they produce. Feed-in Tariffs have been observed to stimulate the engagement of households in the energy system, enabling them to function as both energy producers and consumers in certain countries. Net metering serves as a motivating factor for consumers to engage in the investment of renewable energy generation. Net metering is a mechanism that permits customers to utilize their self-generated electricity to offset their energy consumption. This is achieved by allowing their electric meters to reverse direction, effectively accounting for any surplus electricity generated beyond their immediate needs. In the absence of net metering, it is customary to install an additional meter to quantify the electricity that is fed back into the power grid. The Feed-in Tariff (FiT) is a policy instrument that remunerates electricity producers, ranging from individual households to larger-scale operations, for the power they generate. The forthcoming generation of smart meters intended for deployment in the United Kingdom lacks this particular functionality. The system enables communication with the utility, resulting in improved accuracy and timeliness of billing data. Additionally, it offers the potential for real-time monitoring of household energy consumption (Pollet et al., 2015).

The observed infrastructure advancements indicate a potential shift towards a distinct energy landscape characterized by increased user engagement within the energy system. Nevertheless, the efficacy of these advancements in addressing the interconnected issues of energy security and climate change within the sustainable development framework is a subject of contention. The process of redirecting the trajectory of energy system development is intricate and characterized by a gradual pace. Making decisions regarding the development of nuclear power or carbon capture and storage is a time-consuming endeavor that spans several years. Furthermore, it is important to note that even upon implementation, these solutions will only cater to a fraction of the overall energy system (Frauenfelder et al., 1991).

The issue at hand lacks straightforward resolutions and definitive answers. One of the challenges encountered by policymakers pertains to the absence of a consensus regarding the desired level of greenhouse gas concentration, as well as a definitive schedule for achieving the said target. Determining specific targets and timetables for reaching an agreement remains uncertain and unpredictable. However, it is evident that regardless of the consensus reached, it is imperative to adopt a different perspective when developing the energy system, as the existing structure has proven to be inadequate (Pasqualetti, 2011). Consequently, we must acquire the knowledge and skills necessary to execute tasks markedly and distinctly. It is imperative to critically examine underlying assumptions that form the basis of the existing system. For instance, it is necessary to question whether it is reasonable to anticipate the continuous provision of energy by the system around the clock. Additionally, it is worth considering whether fostering self-sufficiency should be prioritized. Furthermore, deliberation is required to determine whether modifying lifestyles to align with global climate objectives should be an individual responsibility or if it falls within the purview of the government and energy sector (Gallagher et al., 2017).

The answers to these questions are not readily apparent. It is widely acknowledged that energy plays a crucial role in sustaining life and influencing our lifestyles. However, it is imperative to acknowledge the repercussions associated with the generation and utilization of energy, as disregarding these implications in the future would be financially unsustainable. It is imperative to recognize that decisions about energy production and utilization must be approached considering broader societal and environmental implications, necessitating a comprehensive understanding from the individual to the institutional level. The integration of energy system development is imperative for achieving sustainable development (Ballard et al., 2017).

## SUMMARY

This chapter examines the dynamic nature of the energy sector and its possible future directions, considering emerging advancements. The chapter emphasizes several significant trends, namely the promising role of hydrogen as an environmentally friendly energy source, the imperative to decarbonize energy systems, and the importance of emission and carbon trading. The utilization of hydrogen holds promise in sectors that present difficulties in electrification, thereby providing a viable avenue toward the establishment of a hydrogen-centric economy. The influence of government climate policies and regulations on the energy sector is significant, as evidenced by the establishment of renewable energy targets, emission reduction goals, and carbon pricing initiatives. The future trends in energy utilization are subject to constant evolution, influenced by the continuous progress of technology, the development of policies, the dynamics of the market, and the choices made by society.

# REFERENCES

1. Afgan, N. H., & Carvalho, M. G. (2004). Sustainability assessment of hydrogen energy systems. International Journal of Hydrogen Energy, 29(13), 1327-1342.
2. Anke, C. P., Hobbie, H., Schreiber, S., & Möst, D. (2020). Coal phase-outs and carbon prices: Interactions between EU emission trading and national carbon mitigation policies. Energy Policy, 144, 111647.
3. Ballard, A. J., Das, R., Martiniani, S., Mehta, D., Sagun, L., Stevenson, J. D., & Wales, D. J. (2017). Energy landscapes for machine learning. Physical Chemistry Chemical Physics, 19(20), 12585-12603.
4. Blaabjerg, F., & Ionel, D. M. (2015). Renewable energy devices and systems–state-of-the-art technology, research and development, challenges and future trends. Electric Power Components and Systems, 43(12), 1319-1328.
5. Bolla, R., Bruschi, R., Davoli, F., & Cucchietti, F. (2010). Energy efficiency in the future internet: A survey of existing approaches and trends in energy-aware fixed network infrastructures. IEEE Communications Surveys & Tutorials, 13(2), 223-244.
6. Bryngelson, J. D., Onuchic, J. N., Socci, N. D., & Wolynes, P. G. (1995). Funnels, pathways, and the energy landscape of protein folding: a synthesis. Proteins: Structure, Function, and Bioinformatics, 21(3), 167-195.
7. Cadez, S., & Czerny, A. (2016). Climate change mitigation strategies in carbon-intensive firms. Journal of Cleaner Production, 112, 4132-4143.
8. Cames, M., Harthan, R. O., Füssler, J., Lazarus, M., Lee, C. M., Erickson, P., & Spalding-Fecher, R. (2016). How additional is the clean development mechanism. Analysis of the application of current tools and proposed alternatives, 20, 654-656.
9. Chapman, A., Itaoka, K., Hirose, K., Davidson, F. T., Nagasawa, K., Lloyd, A. C., ... & Fujii, Y. (2019). A review of four case studies assessing the potential for hydrogen penetration of the future energy system. International Journal of Hydrogen Energy, 44(13), 6371-6382.
10. Cheng, B., Dai, H., Wang, P., Xie, Y., Chen, L., Zhao, D., & Masui, T. (2016). Impacts of low-carbon power policy on carbon mitigation in Guangdong Province, China. Energy Policy, 88, 515-527.
11. Dawood, F., Anda, M., & Shafiullah, G. M. (2020). Hydrogen production for energy: An overview. International Journal of Hydrogen Energy, 45(7), 3847-3869.

12. de Boer, J., & Zuidema, C. (2015). Towards an integrated energy landscape. Proceedings of the Institution of Civil Engineers-Urban Design and Planning, 168(5), 231-240.
13. De Coninck, H., Haake, F., & Van Der Linden, N. (2007). Technology transfer in the clean development mechanism. Climate policy, 7(5), 444-456.
14. Dechezleprêtre, A., Glachant, M., & Ménière, Y. (2008). The Clean Development Mechanism and the international diffusion of technologies: An empirical study. Energy Policy, 36(4), 1273-1283.
15. Dhanda, K. K., Sarkis, J., & Dhavale, D. G. (2022). Institutional and stakeholder effects on carbon mitigation strategies. Business Strategy and the Environment, 31(3), 782-795.
16. Dhanda, K. K., Sarkis, J., & Dhavale, D. G. (2022). Institutional and stakeholder effects on carbon mitigation strategies. Business Strategy and the Environment, 31(3), 782-795.
17. Dong, H., Geng, Y., Xi, F., & Fujita, T. (2013). Carbon footprint evaluation at industrial park level: A hybrid life cycle assessment approach. Energy Policy, 57, 298-307.
18. Eriksson, E. L. V., & Gray, E. M. (2017). Optimization and integration of hybrid renewable energy hydrogen fuel cell energy systems–A critical review. Applied Energy, 202, 348-364.
19. Fang, G., Lu, L., Tian, L., & Yin, H. (2020). Research on the influence mechanism of carbon trading on new energy—a case study of ESER system for China. Physica A: Statistical Mechanics and its Applications, 545, 123572.
20. Frauenfelder, H., Sligar, S. G., & Wolynes, P. G. (1991). The energy landscapes and motions of proteins. Science, 254(5038), 1598-1603.
21. Gallagher, A. J., Creel, S., Wilson, R. P., & Cooke, S. J. (2017). Energy landscapes and the landscape of fear. Trends in Ecology & Evolution, 32(2), 88-96.
22. Gao, Y., Li, M., Xue, J., & Liu, Y. (2020). Evaluation of effectiveness of China's carbon emissions trading scheme in carbon mitigation. Energy Economics, 90, 104872.
23. Golden, D., Smith, M. A., & Colombo, S. (2011). Forest carbon management and carbon trading: a review of Canadian forest options for climate change mitigation. The Forestry Chronicle, 87(5), 625-635.
24. Hua, W., Jiang, J., Sun, H., & Wu, J. (2020). A blockchain based peer-to-peer trading framework integrating energy and carbon markets. Applied Energy, 279, 115539.

25. Huisingh, D., Zhang, Z., Moore, J. C., Qiao, Q., & Li, Q. (2015). Recent advances in carbon emissions reduction: policies, technologies, monitoring, assessment and modeling. Journal of cleaner production, 103, 1-12.
26. Ke, S., Zhang, Z., & Wang, Y. (2023). China's forest carbon sinks and mitigation potential from carbon sequestration trading perspective. Ecological Indicators, 148, 110054.
27. Li, W., & Jia, Z. (2017). Carbon tax, emission trading, or the mixed policy: which is the most effective strategy for climate change mitigation in China?. Mitigation and Adaptation Strategies for Global Change, 22, 973-992.
28. Lim, X. L., & Lam, W. H. (2014). Review on clean development mechanism (CDM) implementation in Malaysia. Renewable and Sustainable Energy Reviews, 29, 276-285.
29. Long, Y., Li, Y., Wang, Y., Cao, Y., Jiang, L., Zhou, Y., ... & Nakanishi, Y. (2022). Low-carbon economic dispatch considering integrated demand response and multistep carbon trading for multi-energy microgrid. Scientific Reports, 12(1), 6218.
30. Marsden, G., Mullen, C., Bache, I., Bartle, I., & Flinders, M. (2014). Carbon reduction and travel behaviour: Discourses, disputes and contradictions in governance. Transport Policy, 35, 71-78.
31. Midilli, A., Ay, M., Dincer, I., & Rosen, M. A. (2005). On hydrogen and hydrogen energy strategies: I: current status and needs. Renewable and Sustainable Energy Reviews, 9(3), 255-271.
32. Morris, A. J., Meyer, G. J., & Fujita, E. (2009). Molecular approaches to the photocatalytic reduction of carbon dioxide for solar fuels. Accounts of Chemical Research, 42(12), 1983-1994.
33. Nealson, K. H., & Myers, C. R. (1992). Microbial reduction of manganese and iron: new approaches to carbon cycling. Applied and Environmental Microbiology, 58(2), 439-443.
34. Ohndorf, M. (2010). Optimal monitoring for project-based emissions trading systems under incomplete enforcement. Institute for Environmental Decisions Working Paper, (13), 87-90.
35. Onuchic, J. N., Luthey-Schulten, Z., & Wolynes, P. G. (1997). Theory of protein folding: the energy landscape perspective. Annual Review of Physical Chemistry, 48(1), 545-600.
36. Pan, Y., & Dong, F. (2023). Green finance policy coupling effect of fossil energy use rights trading and renewable energy certificates trading on low

carbon economy: Taking China as an example. Economic Analysis and Policy, 77, 658-679.

37. Parra, D., Valverde, L., Pino, F. J., & Patel, M. K. (2019). A review on the role, cost and value of hydrogen energy systems for deep decarbonisation. Renewable and Sustainable Energy Reviews, 101, 279-294.
38. Pasqualetti, M. J. (2011). Social barriers to renewable energy landscapes. Geographical review, 101(2), 201-223.
39. Pasqualetti, M. J. (2012). Reading the changing energy landscape. Sustainable energy landscapes: Designing, planning, and development, 7, 487-489.
40. Patlitzianas, K. D., Flamos, A., Doukas, H., Kagiannas, A. G., & Psarras, J. (2007). Promoting renewable energies and energy efficiency through the CDM funding options. In 2004 New And Renewable Energy Technologies For Sustainable Development (Vol. 2, pp. 1-13).
41. Paul, C., Bartkowski, B., Dönmez, C., Don, A., Mayer, S., Steffens, M., ... & Helming, K. (2023). Carbon farming: Are soil carbon certificates a suitable tool for climate change mitigation?. Journal of Environmental Management, 330, 117142.
42. Pearson, B. (2007). Market failure: why the Clean Development Mechanism won't promote clean development. Journal of Cleaner Production, 15(2), 247-252.
43. Petkov, I., & Gabrielli, P. (2020). Power-to-hydrogen as seasonal energy storage: an uncertainty analysis for optimal design of low-carbon multi-energy systems. Applied Energy, 274, 115197.
44. Pollet, B. G., Staffell, I., & Adamson, K. A. (2015). Current energy landscape in the Republic of South Africa. International Journal of Hydrogen Energy, 40(46), 16685-16701.
45. Quader, M. A., Ahmed, S., Dawal, S. Z., & Nukman, Y. (2016). Present needs, recent progress and future trends of energy-efficient Ultra-Low Carbon Dioxide (CO2) Steelmaking (ULCOS) program. Renewable and Sustainable Energy Reviews, 55, 537-549.
46. Roland, C. M., Santangelo, P. G., & Ngai, K. L. (1999). The application of the energy landscape model to polymers. The Journal of chemical physics, 111(12), 5593-5598.
47. Rosen, M. A., & Koohi-Fayegh, S. (2016). The prospects for hydrogen as an energy carrier: an overview of hydrogen energy and hydrogen energy systems. Energy, Ecology and Environment, 1, 10-29.

48. Sharma, S., Waldman, J., Afshari, S., & Fekete, B. (2019). Status, trends and significance of American hydropower in the changing energy landscape. Renewable and Sustainable Energy Reviews, 101, 112-122.
49. Shojaei, T., & Mokhtar, A. (2022). Carbon mitigation by quota allocation. Journal of Environmental Management, 304, 114097.
50. Song, X., Wang, D., Zhang, X., He, Y., & Wang, Y. (2022). A comparison of the operation of China's carbon trading market and energy market and their spillover effects. Renewable and Sustainable Energy Reviews, 168, 112864.
51. Sørensen, B., Petersen, A. H., Juhl, C., Ravn, H., Søndergren, C., Simonsen, P., ... & Pedersen, T. E. (2004). Hydrogen as an energy carrier: scenarios for future use of hydrogen in the Danish energy system. International Journal of Hydrogen Energy, 29(1), 23-32.
52. Stuhlmacher, M., Patnaik, S., Streletskiy, D., & Taylor, K. (2019). Cap-and-trade and emissions clustering: A spatial-temporal analysis of the European Union Emissions Trading Scheme. Journal of Environmental Management, 249, 109352.
53. Subbarao, S., & Lloyd, B. (2011). Can the clean development mechanism (CDM) deliver?. Energy Policy, 39(3), 1600-1611.
54. Sutter, C., & Parreño, J. C. (2007). Does the current Clean Development Mechanism (CDM) deliver its sustainable development claim? An analysis of officially registered CDM projects. Climatic Change, 84(1), 75-90.
55. Tamalouzt, S., Benyahia, N., Rekioua, T., Rekioua, D., & Abdessemed, R. (2016). Performances analysis of WT-DFIG with PV and fuel cell hybrid power sources system associated with hydrogen storage hybrid energy system. International Journal of Hydrogen Energy, 41(45), 21006-21021.
56. Tang, K., He, C., Ma, C., & Wang, D. (2019). Does carbon farming provide a cost-effective option to mitigate GHG emissions? Evidence from China. Australian Journal of Agricultural and Resource Economics, 63(3), 575-592.
57. Uyar, T. S., & Beşikci, D. (2017). Integration of hydrogen energy systems into renewable energy systems for better design of 100% renewable energy communities. International Journal of Hydrogen Energy, 42(4), 2453-2456.
58. Valentin, J., Dai, A., Nießner, M., Kohli, P., Torr, P., Izadi, S., & Keskin, C. (2016, October). Learning to navigate the energy landscape. In 2016 Fourth International Conference on 3D Vision (3DV). IEEE. (Vol. 2, pp. 323-332).
59. Varanasi, J., & Tripathi, M. M. (2020). Market Models and Operating

Mechanism for Renewable Energy Enabled Indian Electricity Market. Technology and Economics of Smart Grids and Sustainable Energy, 5, 1-9.

60. Vourdoubas, J. (2021). Possibilities for De-carbonizing the Heating and Cooling Sector in the Island of Crete, Greece. European Journal of Environment and Earth Sciences, 2(3), 49-56.
61. Wang, C., Yang, Y., & Zhang, J. (2015). China's sectoral strategies in energy conservation and carbon mitigation. Climate Policy, 15(1), S60-S80.
62. Woerdman, E. (2000). Implementing the Kyoto protocol: why JI and CDM show more promise than international emissions trading. Energy Policy, 28(1), 29-38.
63. Yildiz, B. (2014). "Stretching" the energy landscape of oxides—Effects on electrocatalysis and diffusion. Mrs Bulletin, 39(2), 147-156.
64. Yue, M., Lambert, H., Pahon, E., Roche, R., Jemei, S., & Hissel, D. (2021). Hydrogen energy systems: A critical review of technologies, applications, trends and challenges. Renewable and Sustainable Energy Reviews, 146, 111180.
65. Zainuddin, Z. B., Zailani, S., Govindan, K., Iranmanesh, M., & Amran, A. (2017). Determinants and outcome of a clean development mechanism in Malaysia. Journal of Cleaner Production, 142, 1979-1986.
66. Zerta, M., Schmidt, P. R., Stiller, C., & Landinger, H. (2008). Alternative World Energy Outlook (AWEO) and the role of hydrogen in a changing energy landscape. International Journal of Hydrogen Energy, 33(12), 3021-3025.
67. Zhang, Y., Li, S., Luo, T., & Gao, J. (2020). The effect of emission trading policy on carbon emission reduction: Evidence from an integrated study of pilot regions in China. Journal of Cleaner Production, 265, 121843.
68. Zhang, J., Mauzerall, D. L., Zhu, T., Liang, S., Ezzati, M., & Remais, J. V. (2010). Environmental health in China: progress towards clean air and safe water. The Lancet, 375(9720), 1110-1119.

# INDEX

F



G